CONCEPTS OF MASS

IN CLASSICAL
AND MODERN PHYSICS

CONCEPTS OF MASS

IN CLASSICAL AND MODERN PHYSICS

Max Jammer

1961 | HARVARD UNIVERSITY PRESS
CAMBRIDGE · MASSACHUSETTS

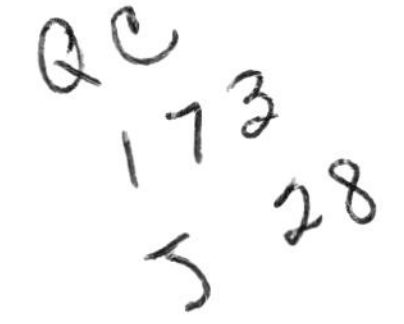

Distributed in Great Britain by Oxford University Press, London

---

Library of Congress Catalog Card Number 61–13737

Printed in the United States of America

RECIPIENT OF THE MONOGRAPH PRIZE
OF THE AMERICAN ACADEMY OF ARTS AND SCIENCES
FOR THE YEAR 1960 IN THE FIELD OF
PHYSICAL AND BIOLOGICAL SCIENCES

PREFACE

This book serves a threefold purpose:

(i) It is designed to present a comprehensive exposition of the historical development of the concept of mass. Although of fundamental importance for physics and the philosophy of science, the notion of mass, apparently, has never heretofore been the subject of an integrated and coherent historical investigation. In spite of its predominantly historical method, the present study puts the emphasis on the intrinsic and conceptual evolution of the notion and treats the purely chronological aspects as of secondary importance.

(ii) The historical research is not regarded as an end in itself. It is certainly true that "one of the most astonishing features of the history of physics is the confusion which surrounds the definition of the key term in dynamics, mass."[1] A critical historical analysis of the classical concepts and definitions of mass and a clear differentiation between inertial and (active and passive) gravitational mass will lead, it is hoped, to a profounder comprehension of the meaning of the term and to a higher level of understanding of its role and significance in physics.

(iii) An absolutely complete elucidation of the concept of mass is admittedly no easy task, for the general and somewhat prob-

[1] G. Burniston Brown, "Gravitational and inertial mass," *American Journal of Physics 28,* 475 (1960).

lematic character of the foundations underlying the exact and unambiguous definition of the concept raises serious questions and involves controversial issues. An adequate presentation of these difficulties is the final purpose of the monograph.

Chapter 4 of the book is an expanded version of a lecture which I gave in Washington, D. C., before the American Association for the Advancement of Science. At a meeting of the Institute for the Unity of Science in Boston, Massachusetts, I had the opportunity of discussing the substance of Chapters 5, 6, and 12. I am grateful to Professor Cyril Stanley Smith (Institute for the Study of Metals, University of Chicago), to Professor Gerald Holton (Department of Physics, Harvard University), and to Professor I. Bernard Cohen (Department of the History of Science, Harvard University) for their valuable comments on these and similar occasions. I am indebted to Professor Adolf Grünbaum (Department of Philosophy, University of Pittsburgh), to Mr. Harry Starr (Lucius N. Littauer Foundation), and to my colleagues at the Department of Physics, Boston University, for their encouragement and advice. My thanks are due to Professor Nathan Rosen (Department of Physics, Technion, Haifa) for his cooperation in the critical reading of the proofs and to Mr. Joseph D. Elder, Science Editor of Harvard University Press, for his careful editing of the manuscript.

MAX JAMMER

Department of Physics
Bar-Ilan University
Ramat-Gan, Israel
April 1961

CONTENTS

INTRODUCTION

What is generally regarded in physics merely as a factor of proportionality, the coefficient of inertia, is the subject of our present investigation.

For the experimentalist engaged in active research the notion of mass seems to offer few difficulties. Since his early training his conceptual tools of thought have constantly been accommodated and conditioned to an uninhibited use of this notion. Conceptual difficulties, if realized at all, are relegated to academic courses at an introductory level. Even in the modern theory of elementary particles and in contemporary field theory, where the notion of mass plays an important role in connection with certain difficulties which at present can hardly be said to have been successfully overcome, it is not the concept as such but its mathematical treatment or presentation that commands the attention of the physicist.

On the other hand, it is generally admitted that for elementary courses in physics the notion of mass is a rather troublesome and intricate subject. Neither textbooks nor lecture courses seem to give a logically as well as scientifically unobjectionable presentation of the concept. In fact, mass is one of those basic concepts whose real significance is fully disclosed and comprehended only gradually in the evolving treatment and progressive discussion of the multifarious phenomena for which it is responsible.

One of the reasons for these difficulties is, of course, the fact that "mass" is a rather abstract concept. It is sometimes claimed that "mass," like "temperature," is closely connected with direct observation or sensory perception. Thus, for instance, Rudolf Carnap in a discussion on the empirical significance of theoretical concepts speaks of a "continuous line from terms which are closely connected with observation, e.g. 'mass' and 'temperature', through more remote terms like 'electromagnetic field' and 'psi-function'."[1] As far as "mass" is concerned, such a characterization — at least with respect to modern physics — seems to be unwarranted. In the eighteenth century, the period of the "substantial conception of matter,"[2] "mass" and "matter" were practically identical and the resistance of matter to pressure, its solidity and impenetrability, or what Leibniz used to call "antitypy," were usually regarded as sensory and directly observable qualities of "mass." The modern concept of mass, however, in contrast to those of "temperature," "light," and "force," has no sensory counterpart nor does it reveal itself directly in any conceivable experiment. It is a construct,[3] in principle to no less an extent than "electromagnetic field." In fact, in modern field theory (as already in the classical electromagnetic theory of matter of Abraham and Lorentz) "field" has methodological priority over "mass" since the attempt is made to describe the latter in terms of the former.

A second difficulty inherent in the concept of mass is the aspect of its metamorphic appearance. "Mass may be compared with an actor who appears on the stage in various disguises, but never as his true self. Actually mass — like the Deity — has a triune personality. It may appear in the role of gravitational charge, or of inertia, or of energy, but nowhere does mass present itself

[1] Rudolf Carnap, "The methodological character of theoretical concepts," in *Minnesota studies in the philosophy of science,* ed. H. Feigl and M. Scriven (University of Minnesota Press, Minneapolis, 1959), vol. 1, p. 39.

[2] This expression is taken from Hermann Weyl, *Philosophy of mathematics and natural science* (Princeton University Press, Princeton, 1949), p. 165.

[3] For this notion see Henry Margenau, *The nature of physical reality* (McGraw-Hill, New York, Toronto, London, 1950), p. 69.

to the senses as its unadorned self."[4] Related to this "multiformity" of "mass" is its "ubiquity," to use another expression from traditional theology, that is, its universal appearance in all branches of physics. This, of course, poses another problem. Arthur Pap in his *Elements of analytical philosophy* alludes to this difficulty when he says: "The fact that a not very refined physicist will in one breath define 'mass' as the tendency or power of a body to resist changes of state (accelerations) and speak of the mass of an electron, is, of course, misleading."[5]

These difficulties and others of a more technical nature justify a comprehensive historico-critical analysis of this all-important concept, especially since no detailed and coherent study of this subject has ever heretofore been published. Although there are naturally certain points of contact with our previous publications on the basic notions of space and force in physics,[6] the present book is a self-contained and independent investigation and does not presuppose a knowledge of the former publications.

It is our firm conviction that a critical analysis of the modern concept of mass leading to a full comprehension of its scientific significance can be presented only through a historical approach. The modern notion of mass had its inception in the seventeenth-century conceptualization of inertial mass primarily by Kepler and Newton. Since Kepler's inertial mass, however, was intimately connected with the preclassical conception of *quantitas materiae* and since the latter notion continued to haunt physical theory for a long time, a comprehensive analysis of our concept on a historical basis should obviously also consider the predecessors and concomitants of our concept.

For the modern mind *quantitas materiae* is at best a meta-

[4] Herbert L. Jackson, "Presentation of the concept of mass to beginning physics students," *American Journal of Physics 27,* 278 (1959).

[5] Arthur Pap, *Elements of analytical philosophy* (Macmillan, New York, 1949), p. 137.

[6] Max Jammer, *Concepts of space* (Harvard University Press, Cambridge, 1954), translated into German as *Das Problem des Raumes* (Wissenschaftliche Buchgesellschaft, Darmstadt, 1960); *Concepts of force* (Harvard University Press, Cambridge, 1957).

physical notion, in contrast to the purely scientific concept of mass; but formerly it was regarded, as its name implies, as a measure of the quantity of matter. The early history of our notion has thus to discuss the general problem of the "how much" of matter or substance, a problem that engaged the attention of many scientists and philosophers.

The classical concept of substance as defined as that which exists in such a manner that it requires no other thing for its existence,[7] when phrased in less metaphysical and more physicalized language, regards matter as the carrier of changing qualities and this carrier itself as unaffected by such changes. Matter is thus something absolute — just as Newtonian space was something absolute as unaffected by material objects,[8] but it is also, in virtue of the very same principle of unaffectedness, something invariant, unchanging, and eternal. Man's quest for preservation and timelessness, which expressed itself in certain religious rituals striving for perpetuation and immortality, found an analogy and ideal in the absoluteness and conservation of matter. It is therefore not surprising that the earliest explicit formulation of the notion of *quantitas materiae* originated in connection with a conceptual analysis of the problem of transubstantiation in the Eucharist. On the other hand, the conception of *quantitas materiae* as a criterion of the "how much" of matter could obviously not fulfill its function. For any measurement — with the possible exception of length determination in classical physics — is ultimately based on an exchange of energy[9] and presupposes an interaction. Something that in principle cannot be acted upon cannot be measured either. Anneliese Maier, to whom we are indebted for her study of Aegidius Romanus and

[7] "Per substantiam nihil aliud intelligere possumus, quam rem quae ita existit, ut nulla alia re indigeat ad existendum," René Descartes, *Principia philosophiae,* I, 51. *Oeuvres de Descartes,* ed. Charles Adam and Paul Tannery (Cerf, Paris, 1905), vol. 8, p. 24.

[8] Cf. Albert Einstein, in the foreword to *Concepts of space,* reference 6.

[9] "Une expérience de mesure comporte toujours . . . une perturbation de ce qu'on veut mesurer." Louis de Broglie.

his concept of *quantitas materiae,* makes the statement that "a real conceptual clarification of *quantitas materiae* or a definition leading to a quantitative operational determination of mass has never been achieved."[10] From our point of view a quantitative determination of matter per se is an impossibility. Matter as such, if in fact there is any need for such a concept in science, necessarily remains an uncomprehended and incomprehensible residuum of scientific analysis and as such unfathomable. Only qualities, carried by matter, so to say, are accessible to quantitative assessment. If quantity of matter is defined through a quantitative aspect of one of its properties, the *quantitas materiae* obviously will depend on what property has been chosen to serve as criterion. In classical physics, for instance, it is a purely accidental feature that two different qualities of matter, inertia and gravitational charge, lead to the same "quantification." Inertia and heat capacity, if chosen as criteria, would have certainly given divergent numerical results.

It is generally stated that "mass" in the sense of inertial mass was introduced into physics by Isaac Newton. As we shall show in detail, the notion of inertial mass was the result of a gradual development which started with Johannes Kepler and terminated with Leonhard Euler. Although the concept of inertial mass is undoubtedly the product of the seventeenth century, its rudimentary sources can be traced back to the Neoplatonic idea of the inertia and inactivity of matter as opposed to the vitality and spontaneity of mind. The ancient metaphysical opposition of matter and spirit was the prototype for the physical antithesis of mass and force. Although Newtonian dynamics prima facie succeeded in alineating the latter concepts from their metaphysical relation, it still left certain conceptual difficulties unsolved.

[10] "Aber trotz all dem ist es nie zu einer wirklichen begrifflichen Klärung der quantitas materiae gekommen, oder gar zu einer Definition, die tatsächlich eine quantitative Erfassung der Masse ermöglicht hätte." Anneliese Maier, *Die Vorläufer Galileis im 14. Jahrhundert* (Storia e Letteratura, Rome, 1941), p. 51.

Modern foundational research, initiated — as far as the concept of mass is concerned — by Saint Venant and by Ernst Mach, tried to resolve these difficulties and achieved important results.

Meanwhile, certain developments in the domain of electrical and optical phenomena gave rise to the concept of field as an elementary category in physics, first as of equal status with that of matter and energy (mass and force) and subsequently as of even a more profound fundamentality than those. Symptomatic for this decline of the mechanical view was the electromagnetic concept of mass, as proclaimed at the turn of the nineteenth century most enthusiastically by Max Abraham, for whom inertia was but an electromagnetic effect. Although this view proved to be untenable, such field-theoretic investigations had a remarkable impact on the development of our notion and led to far-reaching generalizations.

The advent of the special theory of relativity gave rise to a radical modification of the concept of mass and led to an unexpected unification of the previously heterogeneous categories of mass and energy. The three different notions of mass — inertial, active, and passive gravitational mass — which classical physics distinguished conceptually but identified *de facto,* are regarded as equivalent also in the general theory of relativity, although on the basis of completely different considerations. The status of the concept of mass in quantum mechanics and in the physics of elementary particles is still somewhat shrouded in mystery; its clarification is one of the major objectives of the so-called mass-unitarian field theories. A discussion of the role of "mass" in the highly speculative modern space theories of matter concludes the present monograph.

CHAPTER 1

THE ETYMOLOGY OF THE WORD "MASS"

The present chapter is not an integral part of the historico-critical analysis of the concept of mass, for the *word,* and not the *concept,* is its subject. The reader, if he wishes, may skip this chapter without impairing his understanding of the sequel. The chapter discusses the prescientific meaning of the word "mass" or its equivalent in the classical languages, investigates the ancient usage of the term, and speculates on its etymological origin. The author is fully aware of the conjectural character and controversial nature of the chapter.

The word "mass" or its Latin equivalent *massa,* in common use in physics since the beginning of the seventeenth century,[1] was used as a technical term as early as the middle of the fourteenth century by Albert of Saxony in his *Quaestiones super octo libros Physicorum.*[2] Our modern word "mass" (French, *masse;* German, *Masse;* Russian, *масса;* Spanish, *masa*), as used in physics, thus undoubtedly derived from the Latin *massa,* meaning originally a lump of dough or paste. As in the modern languages of today, so already in Middle English the term signified a lump in a more general sense, a conglomeration or

[1] Cf. John Harris (1667?–1719), *Lexicon technicum or an universal English dictionary of arts and sciences,* vol. 1 (London, 1704): "Masse, this word is used by the Natural Philosophers to express the Quantity of Matter in any body."

[2] Part 1, quaestio VI (Venice, 1516; Paris, 1516, 1518).

aggregation of bodies.[3] Such was also the meaning that the word had in the Latin of the Church.[4]

The Vulgate (late fourth century) has *massa caricarum* (1 Regum 25:18), *massa ficorum* (4 Regum 20:7), both in the sense of fig cakes, and *massa palatharum* (2 Regum 16:1), and, without qualifications, "Tulit Daniel picem . . . fecitque massas" (Daniel 14:26).[5] *Massa* in the sense of dough or paste — and obviously unleavened as such — is used in 1 Corinthians 5:6: "modicum fermentum totam massam corrumpit."[6]

Massa in combination with names of metals is often found in Latin, as, for instance, *aurea massa* in *Codex Justinianaeus,* XII, 23, 7, 7 or *Codex Theodosianus,* VI, 30, 7, 7. Similar examples of the use of *massa* are found in Ovid, Vergil, Pliny, Martial, and Juvenal.[7] Isidore of Seville in his *Etymologies*[8] defines *massa* with special reference to metallurgy: "There are three kinds of silver, gold or copper: stamped, wrought, unwrought. Stamped is that in coins, wrought that in vases and statues, unwrought

[3] Cf. for example Maundeville (XIV, 158): "Men fynden . . . hard Dyamandes in a Masse, that cometh out of Gold," quoted from Henry Bradley, *A new English dictionary on historical principles,* ed. Sir J. A. H. Murray (London, 1904–1908), vol. 6, p. 206. Similarly, Ecclus. 22:15, in the 1611 translation of the Bible, reads: "Sand, and salt, and a masse of yron is easier to beare than a man without understanding."

[4] Cf. Quintus Septimus Florens Tertullianus, *De praescriptione haereticorum,* chap. 3, ed. Franciscus Oehler, vol. 2 (Leipzig, 1854): "massa frumenti," p. 5; or Gregorius Turonensis, *De cursu stellarum ratio,* ed. B. Krusch, in *Monumenta germaniae historia,* vol. 1, part 2 (Hanover, 1885), p. 867.

[5] Cf. Prophetia Danielis, *Biblica sacra vulgatae editionis,* ed. Valentinus Loch, vol. 3 (Ratisbonn, 1895), p. 202.

[6] Cf. also Galatians 5:9, and the various meanings of "massa" quoted in Albert Blaise, *Dictionnaire latin-français des auteurs chrétiens* (Strasbourg, 1954).

[7] Ovid, *Metamorphoses,* 11, 112: "contactu glaeba potenti massa fit"; 8, 666: "massa lactis." Vergil, *Georgicon,* 1, 275: "massa picis." Pliny, 36, 6, 8 (49): "marmor, non in columnis crustisve, sed in massa"; 31, 7, 39 (78): "massa salis". Martial, *Epigrammata,* 8, 64, 9: "massa lactis alligati." Juvenal, *Saturae,* 6, 421: "cum lassata gravi ceciderunt bracchia massa," where *massa* is synonymous with *pondus.*

[8] *Sancti Isidori Hispalensis episcopi etymologiarum libri XX,* book 16, chap. 18, *Patrologiae cursus completus,* ed. Migne, vol. 82 (Paris, 1850), p. 585: "Tria autem sunt genera argenti, et auri, et aeris: signatum, factum, infectum. Signatum est quod in numnis est; factum est, quod in vasis et signis; infectum, quod in massis, quod et grave dicitur, id est, massa."

that in masses . . ." It is important to note that the Latin *moles* is often synonymous with *massa,* as for example in Pliny,[9] both terms denoting matter which occupies a certain volume.

There is no doubt that the Latin *massa* is derived from the Greek *maza* (μᾶζα) for "barley cake." Clearly the Latin term has a more general meaning than its Greek original.[10] *Maza,* a common word in Greek literature at the time of Herodotus,[11] designated a kind of bread inferior in quality to the wheaten bread, the *artos.* Aeschylus in his *Agamemnon*[12] uses the term in the particular combination "to eat the bread of slavery." Hippocrates explicitly distinguishes between the more delicate bread (*artos*) and the plain barley cake (*maza*), when he says: "When spring comes . . . substitute for bread barley cake." [13] Aristophanes, in the *Acharnians*[14] and especially in the *Knights,*[15] speaks of *maza* in a most derogatory sense and indicates that, far from being regarded as a delicacy, *maza* was, under the conditions described, not even used for eating but as a spoon for soup after scooping out the soft crumbs. The term was also extensively used by Xenophon, Plato, and Lucian.[16] So much for the Greek original of our modern "mass."

So far our etymological conclusions are well established. When

[9] Pliny, 36, 8, 1.

[10] Cf. A. Ernout and A. Meillet, *Dictionnaire étymologique de la langue latine* (Klincksieck, Paris, 1951), p. 692: "Le mot latin a pris des l'abord un sens plus large que l'original grec, et il en est devenu indépendant . . ."

[11] *Herodoti historiarum libri IX,* ed. H. R. Dietsch, vol. 1 (Leipzig, 1860), p. 107: "Kneading it as it were into a cake" (*hate mazan mazamenos*). The word *maza,* in fact, is found already in the *Epigrammata* ascribed to Homer; see *Homeri opera et reliquiae,* ed. D. M. Monro (Oxford, 1901), p. 1004. See also Hesiod, *Works and days,* verse 819, *Hesiodi carmina, graece et latine,* ed. A. F. Didot (Paris, 1841), p. 42.

[12] Aeschylus, *Agamemnon* 1041, ed. Eduard Fränkel (Oxford, Clarendon Press, 1950), vol. 1, p. 152.

[13] Hippocrates, *Regimen in health,* trans. W. H. S. Jones (Loeb Classical Library; Harvard University Press, Cambridge, 1953), vol. 4, p. 45.

[14] *Aristophanes in English verse,* trans. Arthur S. Way (London, 1927), vol. 1, p. 37.

[15] *Ibid.,* pp. 102, 104.

[16] *Xenophontis De Cyri disciplina libri VIII,* book 6, chap. 2, 28, ed. I. G. Schneider, vol. 1 (Leipzig, 1815), p. 451. Plato, *Republica,* 2, 372b. *Luciani Samosatensis opera,* ed. C. Jacobitz, vol. 1 (Leipzig, 1921), p. 57.

tracing, however, the source of the Greek *maza* we find ourselves confronted with diverging and somewhat contradictory interpretations. Thus Georg Curtius derives the term from the verb *masso* (to knead), with the root *mag-*, and considers the possibility of a relation to *mâc-eria*.[17] Somewhat similarly, Emile Boisacq in his *Dictionnaire étymologique de la langue grecque* derived it from *mag-ia*.[18]

A different theory is proposed by Shipley, who derives the Greek from the Hebrew *mazza* (מצה), the unleavened bread of the Israelites.[19] Eric Partridge[20] also considers the possibility of a Semitic or even Hamitic origin. The first philologist to link the Greek *maza* with the Hebrew was Ernst Assmann. In an article "On the prehistory of Crete"[21] he studied the impact of Semitic (Nabataean and Sabaean) influence on ancient Crete and, through Crete, on Sparta. Pointing out that the constitution of Sparta at the time of Lycurgus followed to some extent the pattern of Crete, Assmann, in particular, discusses the institution of a common meal, the *aiklon* (αἶκλον), on the occasion of which a certain kind of bread, the *maza* (μαζα), was eaten.[22] With respect to the name of this meal Curtius and Meyer[23] agree that its

[17] Georg Curtius, *Grundzüge der griechischen Etymologie* (Leipzig, 1858), part 1, p. 289: "Vielleicht ist auch mâc-eria als geknetete Lehmwand . . . verwandt."

[18] Emile Boisacq, *Dictionnaire étymologique de la langue grecque* (Winter, Universitätsverlag, Heidelberg, 1950), p. 599. See also Osthoff, *Studien zur Lehre von der Wurzelerweiterung und Wurzelvariation* (Upsala, 1891, Upsala universitets arsskrift), pp. 1–294, where it is derived from smē-k-, related to măcer, mager, meaning "triturated."

[19] Joseph T. Shipley, *Dictionary of word origin* (Philosophical Library, New York, ed. 2, 1945), p. 225.

[20] Eric Partridge, *A short etymological dictionary of modern English* (Routledge and Kegan Paul, London, 1958), p. 384.

[21] Ernst Assmann, "Zur Vorgeschichte von Kreta," *Philologus* (Zeitschrift für das klassische Alterthum) *67,* 161–201 (Leipzig, 1908).

[22] "Bei jenem Spartanermahl semitischen Namens gab es auch ein Brot maza, welches recht verdächtig dem hebräischen mazza = ungesäuertes Brot, den Mazzes der Juden gleicht." *Ibid.,* p. 199.

[23] Georg Curtius, reference 17, p. 679; also *Principles of Greek etymology,* ed. 5, vol. 2 (London, 1886), p. 327. Leo Meyer, *Handbuch der griechischen Etymologie,* vol. 1 (Leipzig, 1902), p. 20.

etymological origin lies probably outside the compass of the Indogermanic languages. In fact, *aiklon* stems probably from the Semitic root *akalu,* as for example in the Hebrew *aklah* (to eat). Similarly, Assmann contends, the Greek *maza* is but the Hebrew *mazza.* The possibility of a parallelism in the etymology of the two related terms *aiklon* and *maza* lends, of course, some support to an otherwise controversial issue.

A diametrically opposite hypothesis, as far as the direction of transmission is concerned, has been proposed by Cyrus H. Gordon in his article "Homer and Bible." Gordon writes, with reference to the epic of Kret (Krt: 83–4, 174–5):

> The Kret text records, as worthy of saga, the baking of large bread supplies to equip a large troop movement. This explains the emphasis on the baking of mazza, a non-Semitic word meaning "unleavened bread," in the account of the Exodus from Egypt. The related Greek maza, "barley cake," points to a loan into Hebrew via the Philistines, who may have introduced the Cretan custom as attested in the Cretan Epic of Kret. It is interesting to note that a military institution of the warlike Philistines has been transformed into a cultic phenomenon in Judaism.[24]

The term *mazza* appears frequently, especially in connection with sacrifices, in the Bible: Genesis 19:3; Exodus 12:15, 18, 39; 23:15; 34:18; Leviticus 2:4; 7:12; 8:2, 26; Numbers 6:15; Deuteronomy 16:3, as well as in Joshua 5:11. All these quotations refer to two central ideas. (1) "Seven days shalt thou eat unleavened bread therewith, even the bread of affliction; for thou camest forth out of the land of Egypt in haste."[25] This is a historical reason. In connection with sacrifices, however, a different point is raised: (2) "Thou shalt not offer the blood of my sacrifice with leavened bread."[26] It has been suggested[27] that the prohibition of offering leavened bread at the sacrificial ritual is based on an ancient concept according to which a process of

[24] Cyrus H. Gordon, "Homer and Bible," *Hebrew Union College Annual,* vol. 26 (1955), p. 61. See also *Antiquity, A Quarterly Review of Archaeology 30,* 24–25 (1956).

[25] Deuteronomy 16:3.

[26] Exodus 23:18; similarly, Exodus 34:25, Leviticus 2:4, 11, 7:12.

[27] *Bibel-Lexikon,* ed. Herbert Haag (Benziger Verlag, Zürich, 1956), p. 1097.

fermentation is essentially a process of putrefaction. The leaven thus becomes the symbol of evil influence and moral corruption. Jesus, according to Matthew 16:6, says: "Take heed and beware of the leaven of the Pharisees and of the Sadducees."[28] A still more explicit parallel is drawn in 1 Corinthians 5:7 and 8: "Purge out therefore the old leaven, that ye may be a new lump, as ye are unleavened. For even Christ our passover is sacrificed for us: Therefore let us keep the feast, not with old leaven, neither with the leaven of malice and wickedness; but with the unleavened bread of sincerity and truth."[29]

Additional support for the correctness of this interpretation can be found in the fact that in certain religious institutions of ancient Rome the priests of Jupiter, the so-called *flamines majores Dialis,* were not allowed to touch leaven (*zyme*) or leavened wheaten bread (*aleuron*). Plutarch[30] in his explanation of this prohibition refers to fermentation as putrefaction. This disparagement of leavened bread and the general assumption that it was unleavened bread that was eaten at the Last Supper[31] are perhaps responsible for the fact that the holy bread of the Eucharist, in later centuries, was in general unleavened bread or mazza.

In fact, it has even been suggested that the very name of the Eucharistic service, the "mass," in Old English "maesse," does not originate, as usually contended, from the Latin *missa* (past participle of the verb *mittere,* to send, to dismiss) of the closing phrase of the service: "Ite, missa est." The *New English dictionary on historical principles,*[32] after referring to Isidore of Seville "who conjectured that the original reference was to the

[28] See also Matthew 16:12.

[29] See also 1 Corinthians 5:6.

[30] Plutarch, *Quaestiones romanae,* 109; cf. also *Convivalium disputationum,* 3, 10, 3.

[31] P. Fiebig, in *Theologische Literaturzeitung 59,* 416 (1934), shows that the notion of *artos* as mentioned in the pericope of the Last Supper does not exclude that of *azymon* (unleavened bread) and that consequently the appearance of the term *artos* does not discard the possibility that the Supper was a passover meal.

[32] Reference 3, p. 205.

dismissal of the catechumens which was preliminary to the eucharistic service," adds: "This explanation is not favored by modern scholars . . . Several other theories have been proposed, but none of them has gained wide acceptance among scholars." One of the main arguments against the derivation of "mass" from the "dismissal" (*missio*) of those in the congregation who do not intend to commune lies perhaps in the fact that in cases like this a concrete object — in the present instance, the object sacrificed — usually gives its name to the rite rather than a phrase from the rite itself. In fact, before the sixth century "mass" was called *oblatio* (offering, sacrifice), a term used in Celtic languages to this day (*offeren* in Welsh, Cornish, Irish, Scottish). Similarly, "the sacrifice of the mass" is in present use. The association of "mass," in the sense of religious service, with *mazza,* conjectural as it is, seems nevertheless not completely implausible.

It has also been argued [33] that the institution of the Eucharistic mass as well as its name derive originally from the sacred communion of bread as administered in Mithraism, the serious competitor of Christianity in the Roman Empire. The circularly shaped bread, commemorative of the banquet of Mithras and Helios and bestowing participation in the immortality enjoyed by the god himself, was called *mizd,* a derivative of the older Persian *myazda* which, it is claimed, is the prototype of the Eucharistic *hostia*.[34] If the apostle Paul, as Gardner once argued,[35] was the real originator of the Eucharistic institution, it is not impossible that he was influenced either by the Eleusinian mysteries at Athens or Corinth, or perhaps earlier in Tarsus, his native city, a center of Mithra worship.

[33] See J. A. Magni, "The ethnological background of the Eucharist," *American Journal of Religions, Psychology and Education 4,* 1–47 (1910); also Shipley, reference 19.

[34] Franz Cumont, *Textes et monuments figurés relatifs aux mystères de Mithra* (Brussels, 1896–1899). Heinrich Seel, *Die Mithrageheimnisse während der vor- und urchristlichen Zeit* (Aarau, 1823), derives *mass* or *missa* directly from *mizd.*

[35] Percy Gardner, *The origin of the Lord's Supper* (London, 1893).

Resuming now our etymological investigation, it would now seem appropriate to study the etymology of the Hebrew *mazza*. Unfortunately, however, little reliable information is available. Hebrew philologists, in general, believe that it is derived from *mazzaz* (to press, to suck, to drain), as for instance Gesenius in his well-known *Handwörterbuch zum Alten Testament*.[36] Bäntsch claims that *mazza* is a derivative of the Semitic root *mss* (unripe, tasteless).[37] Recently, Rand advanced the theory that the word originated from the Hebrew root *nazzah,* the equivalent of the Arabic *nada,* implying speed and haste in action.[38]

A different opinion was expressed by Yahuda, for whom the Hebrew *mazza* is "undoubtedly an Egyptian loan-word 'ms·t' or 'msw·t' for a sort of bread or cake and in extended meaning also for food." [39] Finally, Muntner, in an article in the *Krauss jubilee volume,*[40] conjectured that the Hebrew term came from the Assyrian root *mazu* (to press, to pour out, to weaken) or from the Assyrian *macu* (to swell, to extend in space). If the latter hypothesis could be substantiated and the connection between the Hebrew *mazza* and the Greek *maza* irrefutably demonstrated, the final linkage of the physical term "mass" with the ancient notion of spatial extension would exhibit an interesting parallelism to the Arabic term *maddah* (the extended, that which has dimensions) which was used in Muslim philosophy[41] in the sense of "matter." In the context of our discussion it is perhaps not without interest to note that the Arabic translation of the Platonic *ekmageion*[42]

[36] Wilhelm Gesenius, *Hebräisch-Deutsches Handwörterbuch zum Alten Testament* (Leipzig, 1810–1812; ed. 17, Berlin, 1949); English trans. by E. Robinson (Clarendon Press, Oxford, 1957).

[37] Bruno Bäntsch, *Handkommentar zum Alten Testament* (Göttingen, 1903), p. 98.

[38] Abraham Rand, "On the term mazza," *Leshonenu 22,* 81–82 (Hebrew Academy of Language, 1957–58).

[39] A. S. Yahuda, *The language of the Pentateuch in its relation to Egyptian,* vol. 1 (Oxford, 1933), p. 95.

[40] Süssmann Muntner, "Mazza and Maza," *Professor S. Krauss jubilee volume* (R. Mass, Jerusalem, 1937), p. 159.

[41] See, for example, Avicenna, *Shifa',* 1, 6, or A. M. Goichon, *Lexique de la langue philosophique d'Ibn Sina* (Desclée-de Brouwer, Paris, 1938), par. 404.

[42] Plato, *Timaeus* 50 C; *Theatetus* 191 c,d.

(plastic stuff, especially for the impression of seals, mass, in the general sense) was *hamirah,* originally meaning "leavened bread"! [43]

The present philological study, interesting as it may be for itself, was carried out primarily to show that the *word* "mass" originated either from the Greek *maza* (barley-cake or ordinary bread) or possibly from the Hebrew *mazza* (unleavened bread).

In Chapter 4 we shall see how the first explicit definition of the *concept* of mass (as yet in the sense of *quantitas materiae*) originated from a logical analysis concerning the Eucharistic transubstantiation of the Holy Bread.

Thus, the word and the concept, historically speaking, have a common origin.

[43] H. A. Wolfson, "Arabic and Hebrew terms for matter and element," *Jewish Quarterly Review 38,* 47–61 (1947).

CHAPTER 2

DID ANCIENT THOUGHT HAVE A CONCEPT OF MASS?

The physical dimensions that were probably the first to be subjected to the processes of measurement were those of space and time: distance and duration. Already in prehistoric times, however, the rise of intercommunal commerce and the expanding exchange of goods called for ways and means to measure also quantities of goods, such as grain and metals, for whose quantitative determination counting alone was impracticable. Practical necessity thus led to the idea of a quantity of matter, the historical predecessor of our concept of mass.

Two methods were available to antiquity and both were applied: the determination of weight and the determination of volume or bulk.

The use of the balance undoubtedly dates back to prehistoric times. The Osiris religion of ancient Egypt stressed the importance of the balance as a means of assessing and evaluating.[1] In the Bible, Genesis 23:16, we read: "And Abraham weighed to Ephron the silver, which he had named in the audience of the sons of Heth, four hundred shekels of silver, current money with the

[1] *Book of the dead,* chap. 125, *Papyrus Nebseni;* see photographs of the papyrus in the British Museum, Department of Oriental Antiquities, ff. 33, 1876. See also Thomas Ibel, *Die Wage im Altertum und Mittelalter* (Erlangen, 1908) and Lenk, *Die Messkunde als nationales und internationales Problem* (1926).

merchant," a passage which obviously refers to a time when money was still weighed. Coins, as a time-saving device, which made repeated weighing unnecessary, seem to have been introduced in the seventh century B.C.

The earliest standard of measure — apart, of course, from the natural units of time and length (day, month, foot, and so forth) — was interestingly not a unit of weight but a unit of volume: the famous silver vase of Prince Entemena of Lagash (c. 2800 B.C.). Its capacity, as the engraved inscription on its surface indicates, served as a definition of ten *sila* (approximately 5 liters). Units of weight, on the other hand, at that time still varied considerably from place to place and from time to time. Only in the days of Salmanassar the Fifth, king of Assyria (726–722), was the *mina*[2] proclaimed as the official standard of weight (about 1000 grams).

And yet, quantities of different goods were, as a rule, measured in different units.[3] This great variety of ancient units of weight for different kinds of goods — a fact that still reverberates in the metrology of ancient nations in the East — does not find its explanation only in the absence of international agreements at those times. This could not explain the diversity of units in one and the same nation. It is rather the expression of a fundamental philosophy prevailing in ancient thought concerning the concept of weight: weight was not conceived as a dynamical universal quantity or a force, as in modern science, proportional to the quantity of matter or mass (at one and the same locality), but rather as a property of bodies, a quality, like color, odor, or brittleness.

Pierre Boutroux, in an article on the history of the principles of dynamics before Newton,[4] claims that it was probably this

[2] Two different kinds of *mina* were in use at that time, a heavy *mina* and a light one, about half the former.

[3] Hans Joachim von Alberti, *Mass und Gewicht* (Akademieverlag, Berlin, 1957), p. 25.

[4] Pierre Boutroux, "L'histoire des principes de la dynamique avant Newton," *Revue de Métaphysique 28,* 660 (1921).

conception of weight which was responsible for Aristotle's erroneous conclusions concerning the free fall of heavy bodies. The weight of a material particle, according to this conception, depends on whether the particle is part of a big and bulky object or of a small and tiny one. The Peripatetic view that heavy bodies fall faster than light bodies is thus rendered immune to the following charge of inconsistency (which was actually raised in the seventeenth century): two equal, relatively light bodies, regarded as separated from each other, should, according to the Peripatetics, fall with a relatively small velocity; the same bodies, however, viewed as parts of a composite object (of double the weight) should have a great velocity of fall. Weight, in ancient thought, was an intensive rather than an extensive quantity. Within the scheme of Peripatetic thought weight could consequently not be regarded as a measure of quantity of matter with universal applicability to all kinds of material. Furthermore, it could never play the role of a *quantitas materiae,* since such a procedure would necessarily presuppose a correspondence or proportionality between weight and quantity. But such a correlation was absolutely inadmissible, since elements such as fire, or composites of such elements, possessed inherent levity.

For philosophy and science, in contrast to trade and commerce, weight was no measure of the quantity of matter. The question then naturally arises whether according to Aristotle volume or bulk could serve as such a measure. In the *Physics* Aristotle declares: ". . . the matter of a body may also remain identical (*autē*) when it becomes greater or smaller in bulk. This is manifestly the case; for when water is transformed into air the same matter, without taking on anything additional, is transformed from what it was, by passing into the actuality of that which before was only a potentiality to it. And it is just the same when air is transformed into water, the one transition being from smaller to greater bulk, and the other from greater to smaller." [5]

[5] Aristotle, *Physics,* book 4, chap. 9, 217 a 26, trans. P. H. Wicksteed and F. M. Cornford (Loeb Classical Library; Harvard University Press, Cambridge, 1934), vol. 1, pp. 367–369.

Changes of volume do not affect the identity of matter and consequently volume, like weight, cannot serve as a measure of the "quantity of matter."

But could Aristotle really have had a conception of *quantitas materiae*? Could the notion of an invariable "how much" of matter have made sense, even if he had had no operational criterion at his disposal for its quantitative determination? We contend that the notion of *quantitas materiae* as implying a principle of conservation — in spite of its importance in later Scholastic philosophy — is foreign to the original Aristotelian conceptual scheme. In the first place, Aristotelian physics is still the science of *physis*, of the phenomena of growth and becoming; no line of demarcation is drawn between organic and inorganic matter. The very term for matter in the Aristotelian writings, *hylē*, meant originally "wood," "forest" and is probably related to the Indogermanic *sulw* (fertile), much as our modern term "matter" or *materia* (*materies*) meant originally "timber," being related to *mater* (the source of growth).[6] Thus the terminology already indicates the organismic interpretation and conception of the substratum[7] of "physical" phenomena. Now, in organic matter there is growth and decay, increase and decrease of substance; no quantitative permanence or invariance is recognized. This increase or decrease — and this is of crucial importance for our context — in Aristotle's view is conceptually compatible with the preservation of the identity of the substratum and it is in this sense that the last-mentioned quotation from the *Physics* (217 a

[6] That the Aristotelian term *hylē* still carries an organismic connotation can be shown from the use of the term in Homer, *Iliad,* 15:273, 2:455, 23:127, 24:784, 20:491, 11:155, 18:320, 15:606, 16:766, *Odyssey,* 5:470, 17:316, 104, 6:128, 9:234, 14:353; Hesiod, *Works and days,* 420, 511, 490, 1010, 422, 807, 591. Cf. also Aristotle, *Physics,* book 2, chap. 9, 200 b 7; *Metaphysics,* book 8, chap. 6, 1042 a 25. *Materia* or *materies* was used by Cicero, Seneca, and Pliny as a translation for *hylē*. Notker Labeo ("Teutonicus," 950–1022), the early translater of Aristotle's *Hermeneutica,* Boethius's *De consolatione philosophiae,* and Capella's *De nuptiis Philologiae et Mercurii* into Old German, translates *materia* by *zimber,* as does Eckhart (1260–1329).

[7] It still echoes in our term "body" (*corpus, Körper, corps*) as used — somewhat paradoxically — even in the traditional formulation of the law of inertia.

26) has to be understood.[8] The idea that organic growth is a process of addition (absorption, assimilation) of extraneous matter is a mechanistic conception and resulted primarily from the classical experiments on metabolism performed in the seventeenth century by Jan Baptist van Helmont (1577–1644) and Santorio Santorio (1561–1636).

Having convinced ourselves that the concept of mass in the sense of *quantitas materiae* is foreign to Aristotelian thought, we shall ask ourselves whether Aristotle possessed at least an adumbration of the concept of dynamic or inertial mass. Marshall Clagett, in his *Giovanni Marliani and late medieval physics*[9] discusses this question and shows, in opposition to Wohlwill and Duhem, that the answer is negative. The present author fully accepts Clagett's argument. The very fact that for Aristotle movement is the resultant of two forces, one impelling and one resisting but both outside the body itself,[10] seems to settle the question. Aristotle's rejection of the existence of an intrinsic resistance (inertial mass) to an impelling (accelerating) force is implicitly mentioned in *De Caelo,* book 3, chapter 2, 301 b 1–17,[11] where he demonstrates the necessity for every body to have a definite weight (or lightness) on the basis of the fundamental laws of his dynamics. In modern terms it may be said that Aristotle's dynamics is a logically consistent theory for motions either in a gravitational field or in a resisting medium; for motions in a vacuum (whose existence Aristotle does not acknowledge) and in the absence of gravitation his theory breaks down, and this because of the absence of a concept of dynamic mass.

Aristotle's rejection of the possibility of an intrinsic, active resistance of matter to moving force is a consequence of his metaphysical conception of matter according to which its characteris-

[8] See also Aristotle, *Physics,* book 4, chap. 7, 214 a 31.

[9] Columbia University Press, New York, 1941, pp. 125 ff.

[10] Cf. Max Jammer, *Concepts of force* (Harvard University Press, Cambridge, 1957), pp. 39–40.

[11] Aristotle, *On the heavens,* trans. W. K. C. Guthrie (Loeb Classical Library; Harvard University Press, Cambridge, 1939), p. 279.

tics is to suffer action, that is, to be moved.[12] "As to 'matter,' it (qua matter) is passive." [13] For our purpose it is not necessary to discuss the rather complicated Aristotelian theory of prime and secondary matter.[14] One aspect, however, which in the original Aristotelian theory of matter is left as an unsettled problem, deserves elaboration. It is the problem of the relation between matter and magnitude, particularly spatial magnitude, extension. Aristotle repeatedly describes "matter" as "extended body" (*soma*). *Soma,* in conformity with our previous discussion of the organismic character of the Aristotelian physics, denoted only the human body, either dead (cadaver) as in *Iliad* 3:23 and 7:79, or alive, as in Hesiod's *Works and days* 540.[15] Aristotle uses the term in accordance with the definition given in *Physics,* book 3, chapter 5, 204 b 6: ". . . 'body' is defined as 'that, the limit or boundary of which is surface'," [16] that is, in the sense of a geometric three-dimensional volume.

On the other hand, numerous statements in Aristotle affirm that the first matter is not body and has no magnitude, as for example in *Metaphysics,* book 7, chapter 3, 1029 a 20: "By matter I mean that which in itself is neither a particular thing nor of a certain quantity nor assigned to any other of the categories by which being is determined." [17] Simplicius was puzzled by this apparent contradiction. "If matter is body," he says, "it must be a certain quantum (*poson*) and endued with magnitude (*memegethysmenon*)." [18] Furthermore, Aristotle's deduction of

[12] Aristotle, *De generatione et corruptione,* book 2, chap. 9, 335 b 30.

[13] *Ibid.,* 324 b 18.

[14] See, for example, Augustin Mansion, *Introduction à la physique Aristotélicienne* (Vrin, Louvain, Paris, 1945), chap. 5.

[15] Only with Empedocles and the atomists does the term acquire its later universal meaning.

[16] Aristotle, *Physics,* book 3, chap. 5, 204 b 6 (Loeb Classical Library), vol. 1, p. 233. See also *De caelo,* book 1, chap. 1, 268 a 1–10; *Metaphysics,* book 5, chap. 6, 1016 b 24–28; *Topics,* book 4, chap. 5, 142 b 24.

[17] "Dico autem materiam, quae per se ipsam neque quid neque quantum nec aliud quippiam dicitur, quibus ens determinatur."

[18] *Simplicii in Aristotelis Physicorum libros quattuor priores commentaria,* ed. Hermann Diels (Berlin, 1882), p. 229.

the existence of matter from the transmutation of the four elements, as the common substratum of the interchangeable qualities of hot, cold, moist, and dry, proves to Simplicius that Aristotelian matter is corporeal and extended. In an attempt to solve this dilemma he suggests that the first matter is inextended per se but through the intermediation of a corporeal form becomes extended matter and thus the underlying substratum of the four elements.

May we not, therefore, admit that body is twofold, one kind, as subsisting according to form and reason, and as defined by the intervals; but another as characterized by intensions and remissions, and an indefiniteness of an incorporeal, impartible, and intelligible nature; this is not being formally defined by three intervals, but entirely remitted and dissipated, and on all sides flowing from being into nonbeing. Such an interval as this, we must, perhaps, admit matter to be, and not corporeal form, which now measures and bounds the infinite and indefinite nature of such an interval as this, and which stops it in its flight from being.[19]

Simplicius' critique of Aristotle's theory of matter has been discussed at some length because of its implantation of the notion of corporeal form as the introducer of quantity, an idea of considerable importance for medieval philosophy. It is also important to note that Simplicius' innovation characterizes spatial extension as the first and foremost measurable property of matter. Like Simplicius, later Peripatetic philosophers accepted spatial extension as the quantitative measure of matter. It was essentially Euclid's definition 1 in book 11 of his *Elements* which was taken as applicable not only for geometry but for physics as well: "A solid is that which has length, breadth, and depth." [20]

The question, of course, arises whether this definition, in fact, was meaningful also for physics. As far as the comparison of different quantities of one and the same homogeneous material is concerned, it is clear that it was. But rarely was there a need to compare different substances. Impact or collision experiments, centrifugal effects, and local variations of gravitational attrac-

[19] *Ibid.*, p. 230.
[20] "Solidum est quod longitudinem et latitudinem et altitudinem habet."

tion, three phenomena that could reveal the insufficiency of this approach, and that actually gave rise to the dynamical concept of mass, were completely outside the scope of ancient and medieval science.

If for the Peripatetics spatial extension could serve as a measure for the quantity of matter, Platonists and Neoplatonists certainly faced no difficulty in this respect. In Plato's identification of physical bodies with the world of geometric forms, in his fundamental idea of a geometrization of physics, geometric extension is the one and uniform invariant which remains always and everywhere the same.[21]

> Suppose a person to make all kinds of figures of gold and to be always transmuting one form into all the rest; — somebody points to one of them and asks what it is. By far the safest and truest answer is, That is gold; and not to call the triangle or any other figures which are formed in the gold "these," as though they had existed, since they are in process of change while he is making the assertion . . . The same argument applies to the universal nature which receives all bodies — that must be always called the same; for while receiving all things, she never departs at all from her own nature, and never in any way, or at any time, assumes a form like that of any of the things which enter into her; she is the natural recipient of all impressions.[22]

Space as the matrix of all things is permanent and thus for Plato a reliable quantitative indicator.

The greatest opposition, in antiquity, to these ideas was voiced by the Stoics, who stressed the distinction between space and body. But what really is the difference between these two? Clearly, they claim, body is more than mathematical extension, physics more than geometry. Do we not get the notion of geometric extension by *abstraction* from physical bodies? The answer was found in *antitypia,* the resistance of a body to mechanical pressure which according to the Stoics prevents total blending or

[21] J. Bassfreund, *Über das Prinzip des Sinnlichen oder die Materie bei Plato* (Leipzig, 1886), p. 48; J. S. Könitzer, *Über Verhältnis, Form und Wesen der Elementarkörper in Platos Timaeus* (Neu-Ruppin, 1846).

[22] Plato, *Timaeus,* 50 a, *The dialogues of Plato,* trans. B. Jowett (Random House, New York, 1937), vol. 2, p. 30.

intermixture (*krasis*) of all elements.[23] The Skeptic Sextus Empiricus repeatedly characterizes a physical body as that which possesses magnitude, shape, resistance, and weight.[24] The first two attributes apply to a body qua geometric extension and the other two attributes make the geometric entity a physical body. However, in all these speculations neither weight nor resistance is ever considered what we would call an extensive quantity, capable of being used as a measure, as a *quantum materiae*. These attributes, magnitude and shape included, are for these ancient philosophers, as Plotinus repeatedly explains,[25] rather forms and not the substratum that accepts the forms. Elementary matter, matter per se, cannot be described in a quantitative manner — in diametric opposition to Newtonian and modern physics, which finds its task in reducing qualities to quantitative aspects and which does not regard quantity as a special kind of quality (or form). This fundamental difference in the conceptual outlook of ancient thought from that of modern science explains the absence of the notion of *quantitas materiae* in antiquity, whereas in modern science the quantity of inertial mass — the modern measure of *quantum materiae* — plays such an important role.

On the other hand, one may claim that the ancient concept of quantification of matter as the application of a form (*forma*) is somewhat analogous to the view of the modern philosophy of physics. Every measurement, as we know today, implies an interaction between the object to be measured on the one hand and the recording apparatus or instrument on the other. Length, duration, or mass in modern physics, as against classical (Newtonian) physics, are not intrinsic properties of the object under discussion, but result from certain physical operations. These opera-

[23] Cf. Plotinus, *Enneads,* 6, 1, 26; Sextus Empiricus, *Against the mathematicians,* 10, 221; Plutarch, *Against Colotes,* chap. 16, 1116 D. Epicurus assigned *antitypia* to matter as against *eixis* (resistlessness) of the void.

[24] Sextus Empiricus, *Against the mathematicians,* 1, 21; 10, 240, 257; 11, 226.

[25] Plotinus, *Enneads,* 3, 6, 17; 2, 4, 8.

tions are interactions, either strong ones, as in the determination of the momentum of an elementary particle, or weak ones, as in the determination of the temperature of a macroscopic body. Yet every physical quantity, to be measurable, must give rise to some physical activity, some exchange of energy. Newtonian physics also admitted essentially this idea; it demanded, however, two important provisions: (1) the effect exerted by the measuring instrument upon the object of measurement can be made (in principle at least) arbitrarily small and can be corrected for; (2) by regarding spatial extension (and temporal duration) as a purely geometric, essentially nonphysical quantity, the determination of length or volume (as well as time intervals) was considered to be exempted from such an interaction. Quantum mechanics and relativity in modern physics invalidated these restrictions. Now if matter is conceived as an absolutely passive, inert, and in every respect inactive substratum of the material world, the concept of a "quantity of matter" in the sense of a measurable characteristic of a physical object becomes a *contradictio in adjecto*. The passivity of matter precludes its quantification. Quantification of matter, for Plotinus and, as we saw above, for Simplicius, is the result of an introduction of form; for modern physics it is the result of an operation.

The absence of a conceptualization of "quantity of matter" in ancient philosophy and science — apart from the practical methods of volume and weight determinations — was not regarded as inconsistent with the general assertion of a principle of conservation of matter. The question of an experimental verification of such a principle, which would have raised the problem of consistency, was never asked since the principle was considered only in its cosmologic aspects.[26] The principle of the permanence of matter was an essential foundation of Democritus' metaphysics:

[26] Aristotle discusses the principle only *en passant* in a rather obscure passage in *Physics,* book 1, chap. 9, 192 a 26–34, which puzzled Simplicius; and in 191 b 30 he maintains the principle "ex nihilo nihil fit" for the absolutely nonexistent but not with respect to the incidentally nonexistent.

"Out of nothing nothing comes and nothing turns into nothing."[27] Similarly, Plutarch quoted Empedocles as saying:

> Fools, and of little thought, we well may deem
> Those, who so silly are as to esteem
> That what ne'er was may now engendered be,
> And that what is may perish utterly.[28]

An often quoted expression of the principle of the uncreatedness of matter is Lucretius's famous statement which serves as the starting point in his great philosophical poem *On the nature of things:* "Nothing can ever be created out of nothing by divine power."[29] As his second great principle he affirms the indestructibility of matter: "Nature resolves everything into its component atoms and never reduces anything to nothing."[30] Applying these principles to his doctrine of atoms and vacuities he asks: "Why do we find some things outweigh others of equal volume? If there is as much matter in a ball of wool as in one of lead, it is natural that it should weigh as heavily, since it is the function of matter to press everything downwards, while it is the function of space on the other hand to remain weightless."[31] These passages show clearly that Lucretius's conceptual scheme is fundamentally different from the Aristotelian pattern of thought. In the first place, in accordance with the assumption[32] that all atoms are endowed with weight, weight is no longer an accidental property of matter but becomes a universal attribute of matter. Secondly, the passage

[27] Democritus, in Diogenes Laertius's *De clarorum philosophorum vitiis,* ed. G. Cobet (Paris, 1850), book 9, 44, p. 238: "Nihil ex eo quod non sit fieri, neque in id quod haudquaquam sit corrumpi." Aristotle ascribes this idea to Anaxagoras; see *Physics,* book 1, chap. 4, 187 a 34: "Nothing can come out of what does not exist."

[28] Plutarch, "Against Colotes," *Moralia,* ed. W. W. Goodwin (Boston, 1870), vol. 5, p. 351.

[29] T. Lucretius Carus, *De rerum natura,* book 1, verse 150: "nullam rem e nilo gigni divinitus umquam."

[30] Lucretius, *The nature of the universe,* trans. R. E. Latham (Penguin Classics, Melbourne, London, Baltimore, 1952), p. 33.

[31] *De rerum natura,* book 1, verse 360: "nam si tantundemst in lanae glomera quantum corporis in plumbo est, tantundem pendere par est."

[32] Modern scholarship does not universally accept the contention that ancient atomism ascribed weight to atoms as a universal property.

clearly suggests a proportionality between "as much matter" (*quantum corporis*) on the one hand and "weight" (*pendere*) on the other. Thus for Lucretius weight could have served as a measure of the quantity of matter, and the principle of indestructibility of matter could have been given an operational interpretation as the principle of conservation of weight.

That such an idea was not completely unthought of can be gathered from the writings of Lucian of Samosata (A.D. 125–180). In his *Life of Demonax* (one of the Cynics of the second century A.D.), Lucian describes the biting wit of Demonax: "Even for questions meant to be insoluble he generally had a shrewd answer at command. Some one tried to make a fool of him by asking, 'If I burn a hundred pounds of wood, how many pounds of smoke shall I get?' 'Weigh the ashes, the difference is all smoke.' "[33] Such ideas, however, remained isolated statements. Never, in antiquity, was the conservation of weight explicitly expressed as a scientific principle. And never did such ideas give rise to the formation of the concept of "quantity of matter" in a technical sense.

One may argue, perhaps, that the experiments and theories in hydrostatics, especially those considered by Archimedes, should have led — via the notions of specific gravity and density — to the concept of mass. As far as the notions of density and rarity are concerned, it can be shown that they were treated primarily as irreducible qualities, much on the same footing as color, odor, or other accidents of matter. Gotifredus, Marsilius, and Pompanatius still contended in their scholastic analyses of the processes of condensation and rarefaction that in these phenomena the former accidents corrupt and new accidents arise, whereas the Scotists, Hervaeus, and Buridan insist that a gradual accumulation of new accidents takes place.[34] One should not be misled by Scholastic statements such as "in raro parvum est de materia sub magnis dimensionibus, et in denso multum sub parvis." The no-

[33] *The works of Lucian of Samosata,* trans. H. W. Fowler and F. G. Fowler (Oxford University Press, London, 1905), vol. 3, p. 9.

[34] See Antonius Rubius, *De generatione et corruptione* (commentary) (Lyons: 1614), book 1, chap. 5, quaest. 2, number 26, 27.

tion of density in its exact definition as the ratio of mass to volume is a rather recent concept, dating back to Leonhard Euler. The notion of specific gravity, referred to the standard of water, is much older;[35] it is mentioned explicitly in the thirteenth century pseudo-Archimedean treatise *De ponderibus* (or *De insidentibus in humidum*), but was employed already by Abd al-Rahman al-Manzur Al-Khazini (fl. c. 1120) in his *Book of the balance of wisdom*[36] and still earlier by Muhammad ibn-Ahmad Al-Biruni (873–1048) in his *Ayin-Akbari*.[37]

Although the idea of specific gravity appears implicitly also in Archimedes' treatises on hydrostatics,[38] the term is never used or defined by Archimedes. Modern writers of textbooks and popularizers of science often credit Archimedes not only with the notion of specific gravity, but also with that of density. This error may have been caused by the following philologic inaccuracy. The Greek text of Archimedes' treatise *On floating bodies* had been known only since 1899. Till then, Latin translations were in use, such as those of Moerbeke (sometimes ascribed to Tartaglia), of Barrow, or of Torelli.[39] In these translations the Archimedean term *onkos* (volume) was rendered as *moles*, which in turn was

[35] K. B. Hofmann, "Kenntnisse der klassischen Völker von den physikalischen Eigenschaften des Wassers," *Sitzungsberichte der kaiserlichen Akademie der Wissenschaften in Wien 163*, 17 (1909); E. O. von Lippmann, *Abhandlungen und Vorträge zur Geschichte der Naturwissenschaften*, Abh. 2 (Leipzig, 1913), p. 175; H. Schelenz, "Zur Geschichte der Volumgewichts-Ermittlung," *Chemiker Zeitung 39*, 913 (1915).

[36] *Kitab mizan al-hikma* (1121–1122). Cf. N. Khanikoff, "Analysis and extracts of the Book of the balance of wisdom," *Journal of the American Oriental Society 6*, 1–128 (1859).

[37] J. J. Clement-Mullet, "Pesanteur spécifique de diverses substances minérales" (excerpts from *Ayin-Akbari*), *Journal Asiatique 11*, 379–406 (1858). Cf. also Eilhard Wiedemann, "Über die Kenntnisse der Muslime auf dem Gebiete der Mechanik und Hydrostatik," *Archiv für die Geschichte der Naturwissenschaften 2*, 394–398 (1910).

[38] Archimedes, *On floating bodies;* cf. E. J. Dijksterhuis, "Archimedes," *Acta historica scientiarum naturalium et medicinalium 12*, 375 (Munksgaard, Copenhagen, 1956).

[39] William of Moerbeke (c. 1215-c. 1286), *De iis quae in humido vehuntur* (Bologna, 1565); Isaac Barrow, *Archimedis opera* (London, 1675), p. 246; J. Torelli, *Archimedis quae supersunt omnia* (Oxford, 1792).

interpreted as "mass." Another possible reason for this error is the fact that many textbooks of physics discuss the principle of Archimedes immediately after having treated the fundamental concepts of mechanics and, among these, the notion of mass.[40] Vitruvius, in his treatise *On architecture*,[41] relates the well-known story of how Archimedes established his hydrostatic principles: ". . . He is said to have taken two masses (*duas massas*) of the same weight as the crown . . ." *Massa,* in this context, stands for "lump," "block" and should not be taken as a technical term in the modern sense of the word. This may be an additional reason for erroneously associating Archimedes with the concepts of density and mass.

We thus see that the ancient science of hydrostatics, its concepts and its terminology, cannot disprove our conclusion that antiquity did not formulate a concept of mass, either in the sense of *quantitas materiae* or in the sense of dynamic mass.

[40] For instance, C. E. Mendenhall, A. S. Eve, D. A. Keys, and R. M. Sutton, *College physics* (Heath, Boston, 1944), p. 116, opens the discussion of Archimedes' principle with the following words: "Let a stone of mass *m* be attached to a spring balance . . ."

[41] Marcus Vitruvius Pollio, *De architectura,* trans. F. Granger (Loeb Classical Library; Harvard University Press, Cambridge, 1934), vol. 2, pp. 204–205.

CHAPTER 3

THE NEOPLATONIC CONCEPT OF INERTIA

In the philosophy of Plato and his successors it was ultimately the Pythagorean element that led to the geometrization of physics, to the metaphysical identification of matter with space, and thus to the theoretical possibility of a quantitative determination of matter via the determination of volume or bulk. For scientific and technologic applications, however, these conclusions proved impracticable. The future development of science could not assimilate these ideas and could not adopt them, in particular, as a basis for a concept of mass.

Another line of thought that emerged from the amalgamation of Platonic with Judeo-Christian philosophy in the early Middle Ages proved to be of greater importance for the future development of our concept. Historically viewed, it is a curious and somewhat paradoxical development that the very emphasis of Neoplatonic and Judeo-Christian thought upon the essentially spiritual and immaterial nature of reality gave rise to an idea which in the future development of scientific thought was destined to form the basis of a materialistic, substantial philosophy. In their effort to show that all force and life have their source in the intellect and in God, Neoplatonism and Judeo-Christian philosophy degraded matter to impotence and endowed it with "inertia" in the sense of an absolute absence of spontaneous activity or "form." The Aristotelian idea of privation, in Aristotle still a neutral and indifferent concept, now became a predicate of depravation and debasement. But it was this very notion of inertia which with the rise of classical mechanics in the seventeenth century, and after being gradually purged of its derogatory emotional connotations, became the characteristic criterion for the dynamic behavior of matter and thus the foundation for the concept of inertial mass.

In contrast to the Stoa, Plotinus, the early founder of Neoplatonism, distinguishes between matter and body (corporality); body, for him, is a composite of which matter is an element.[1] As an incorporeal existent, matter is only "an image and phantom of extension." Nature, usually regarded as the activity of matter, is actually that which when added to matter gives it substantiality. A typical Neoplatonic appraisal of matter is given by Plotinus in the following passage:

> Matter is a fugitive bauble and so are the things that appear to be in it, mere shadows in a shadow. As in a mirror the semblance is in one place, the substance in another, so matter seems to be full when it is empty, and contains nothing while seeming to contain all things. The copies and shadows of real things which pass in and out of it come into it as into a formless shadow. They are seen in it because it has no form of its own; they seem to act upon it, but they produce nothing, for they are feeble and weak and have no power of resistance (*soliditatem*). But neither has matter any such power; so they go through it like water without clearing a passage."[2]

Matter is still the Platonic matrix of all forms. But a new idea is already added: the material substratum of physical existence, characterized by its passivity alone, becomes endowed with extension through the intervention of a substantial form, an idea whose (later) occurrence in Simplicius has already been mentioned. Proclus, the other great exponent of Neoplatonism in the East, accepts Plotinus's doctrine, but with one important modification: the passivity or inertia of matter follows from its spatial

[1] Plotinus, *Enneads* 3, book 6, chap. 7: "Est igitur (materia) incorporea, quoniam corpus posterius est atque compositum, et ipsa una cum alio complet corpus," *Plotini opera omnia,* ed. G. F. Creutzer (Oxford, 1835), p. 566.

[2] *Ibid.*

extension. "For matter, as matter (*soma*), has no character save divisibility, which renders it capable of being acted upon, being in every part subject to division, and that to infinity in every part."[3] Since extension, in Proclus's view, is tantamount to unlimited divisibility, something infinitely divisible is subject to external activity to an unlimited extent and consequently is only of a passive nature. Thus, according to Proclus, extension qua divisibility gives rise to passivity or inertia.

The inference from spatial extension to inertia — always, of course, in the sense of absolute passivity — found a more explicit expression in the philosophy of the so-called Brethren of Purity (Ihwan-es-Safa), a secret religious, philosophical, and political association established at Basra at the end of the tenth century. Their philosophy was most probably influenced by Neoplatonic thought. Excerpts from the writings of Plotinus and Proclus, as we know today, reached the Muslims as pseudoepigraphic writings in an Arabic translation in the middle of the ninth century[4] and contributed to a neoplatonization of Aristotle in Muslim philosophy. The Platonic identification of matter with spatial extension is fully accepted by the Brethren of Purity and becomes the attribute of matter that characterizes its essence. "The philosophers say: 'Body is the thing that has length, breadth, and depth. Thing, here, refers to matter, i.e. substance, and length, breadth, and depth to form, by which matter becomes a three-dimensional body."[5] The important innovation, however, is another statement: "Rest is more in accord with the

[3] Proclus, *The elements of theology,* ed. E. R. Dodds (Oxford University Press, 1933), p. 75.

[4] Plotinus's *Enneads,* books 4–6 (in part) were translated into Arabic by Na'imah ibn Abdallah Abdal-Mesih in about 840 and circulated under the title *Theology of Aristotle.* Proclus's *Institutio theologica* (in part) forms the core of the *Liber de causis* which was translated into Arabic in 850 approximately.

[5] See Friedrich Dieterici, *Die Lehre von der Weltseele,* in *Die Philosophie der Araber im 9. und 10. Jahrhundert,* book 8 (Leipzig, 1872), p. 127. Cf. also in the same series *Die Naturanschauung und Naturphilosophie der Araber,* book 5 (Leipzig, 1876), p. 3, and *Die Anthropologie,* book 7 (Leipzig, 1871), p. 21.

concept of matter than motion, because matter, though having six sides (i.e. three dimensions), cannot move simultaneously in all six directions; and a movement in one direction is not preferential to that in another; therefore immobility is characteristic of matter." The inertia of matter, qua immobility, is thus, by the principle of sufficient reason, a consequence of its spatial extension.

The same conclusion, though by a different argument, is drawn by the eleventh-century Spanish-Jewish philosopher Ibn Gabirol (Avencebrol), who is generally regarded as the first exponent of Neoplatonism in the West. Although, according to Munk,[6] Ibn Gabirol apparently had no direct knowledge of the writings of Plotinus and Proclus, their influence on his system of thought cannot be denied.[7] For Ibn Gabirol quantity, and spatial extension in particular, opposes and forecloses all activity. In his *Fons vitae* he distinguishes between corporeality and materiality, but contrary to Plotinus, for whom matter occupies the lowest scale in the gradation of existence, Ibn Gabirol conceives matter as the underlying substance of all beings, with the sole exception of God himself. The further matter extends from its highest gradation, the spirit, the more corporeal it becomes. The lowest kind of matter, the corporeal world, is characterized by spatial extension and is inert. "Quantity prevents substance from conferring its essence, because of its bulkiness and circumference."[8] "Just as moisture prevents the spread of fire and clouds the penetration of light, so extension precludes any activity of matter."[9] This correlation between extension and inactivity, inertia and im-

[6] Salomon Munk, *Mélanges de philosophie juive et arabe* (Paris, 1859), p. 240.

[7] M. Joël, "Ibn Gabirol's Bedeutung für die Geschichte der Philosophie," *Monatsschrift für die Geschichte und Wissenschaft des Judentums 6* (1857), reprinted in M. Joël, *Beiträge zur Geschichte der Philosophie* (Breslau, 1876); investigates Gabirol's relation to Plotinus.

[8] "Quantitas est prohibens ne substantia conferat suam essentiam, propter crassitudinem quantitatis et suam circumscriptionem." *Fons vitae,* ed. Clemens Bäumker, *Beiträge zur Geschichte der Philosophie des Mittelalters,* vol. 3 (Münster, 1895), p. 198.

[9] *Ibid.,* vol. 2, p. 41.

mobility, reveals itself, according to Ibn Gabirol, in the simple everyday experience: the greater the spatial extension, the heavier the object; the heavier the object, the less mobile it is.[10] For Ibn Gabirol's philosophy of emanations the inertia of corporeal matter was a logical conclusion of its lowest gradation in the hierarchy of being. For every activity, to be carried out, necessitates something subjacent, something of a lower rank, to be acted *upon.* Since corporeal matter is the existent of lowest rank, it can have no substratum on which to act and thus remains necessarily inert.

This degradation of matter as the principle of inactivity and inertia is found earlier in Plotinus[11] and especially in Philo of Alexandria. For Philo, matter is even the principle of evil (*cheirōn ousia*),[12] it is unclean and thus could never have been in immediate contact with God. This deprecatory attitude toward matter explains why Philo Judaeus accepted the Platonic, but wholly un-Biblical and un-Jewish, concept of the uncreatedness of matter.[13]

Another disparagement of matter, related to Philo's contemptuous view, is the idea of the ugliness of matter. The earliest expression of such an idea is perhaps found in Plotinus, for whom matter as the receptacle of form is necessarily itself formless; since form alone can give beauty, matter is ugly, not in the sense of being deformed or disfigured but rather in the sense of being shapeless and clumsy.[14]

Although Chalcidius's *Commentary to Timaeus,* which had a predominant influence on the early philosophy in the West, propounded a strictly Aristotelian theory of matter, the doctrine of

[10] "Discipulus: Quod signum est quod quantitas substantiam in qua subsistit retinet a motu? Magister: Signum huius rei invenies in re manifesta, quia omne corpus, quanto magis accreverit eius quantitas, erit gravius ad movendum et ponderosius . . . quantitas est causa efficiens ponderositatis et prohibens a motu." *Ibid.*

[11] Plotinus, *Enneads,* 1, 8, 3; 6, 7, 28.

[12] *Philonis Judaei opera omnia* (Leipzig, 1828), *De creatione principum,* 7.

[13] Philo, *De sacrificiis Abelis et Caim.*

[14] Plotinus, *Enneads,* 1, 6, 2.

matter in early patristic and scholastic philosophy was primarily a fusion of Neoplatonic (mostly Plotinistic) and Christian-theological elements. Throughout medieval thought, matter is conceived in the Plotinistic-Philonistic way as an inert, shapeless, clumsy being, as a detailed study of the Venerable Bede,[15] Alcuin,[16] Rhabanus Maurus,[17] Remigius of Auxerre,[18] Anselm of Canterbury,[19] Honorius Augustodunensis,[20] Hugo of St. Victor,[21] Robertus Pullus,[22] Petrus Abaelardus,[23] Petrus Lombardus,[24] Petrus Pictaviensis,[25] Petrus Comestor,[26] and William of Conches[27] will show.

It would certainly lead us far astray to discuss in detail the continuous transmission of the Plotinistic concept of matter throughout this voluminous literature. Two examples, not listed in the foregoing references, will suffice. In a twelfth-century treatise, entitled *De mundi universitate libri duo,*[28] ascribed to Bernard of Chartres but written probably by Bernard Silvestris, matter is described as a shapeless, inert, and confused mass of elements; it is both the monstrous chaos of the Greeks and the "earth without form and void" of the Bible.[29] Alanus ab Insulis, a thirteenth-century author, in his encyclopedic poem *Anticlaudian*

[15] *Hexaemeron,* book 1, *Opera omnia,* J. P. Migne, *Patrologia latina,* vol. 91, p. 15 A, B.

[16] *Interrogationes et responsiones in Genesin,* Migne, vol. 100, p. 519 B.

[17] *Commentaria in Genesin,* book 1, Migne, vol. 107, p. 446 B.

[18] *Commentarius in Genesim,* part 1, Migne, vol. 131, p. 55 A, B.

[19] *Monologium,* chap. 7, Migne, vol. 158, p. 153 C, D.

[20] *Hexaemeron,* chap. 1, Migne, vol. 172, p. 225 A.

[21] *Adnotationes elucidatoriae in Pentateuchon,* chap. 5, Migne, vol. 175, p. 34 B, C.

[22] *Sententiarum libri octo,* book 2, chap. 1, Migne, vol. 186, p. 717 D.

[23] *Expositio in Hexaemeron,* Migne, vol. 178, pp. 733 C, 735 A.

[24] *Sententiarum libri quatuor,* book 2, Migne, vol. 192, p. 675.

[25] *Sententiarum libri quinque,* book 2, chap. 7, Migne, vol. 211, p. 958 C.

[26] *Historia libri Genesis* (Historia scholastica, chap. 1), Migne, vol. 198, p. 1055 B.

[27] *Elementa philosophiae,* chap. 1.

[28] Book 2, chap. 1, verse 5: "Ecce, inquit, mundus, o natura, quem de antiquo seminario, quem vultu vetui, quem de massa confusionis excepi." *Bibliotheca philosophorum mediae aetatis,* ed. C. S. Barach and J. Wrobel (Innsbruck, 1876), p. 33.

[29] Genesis 1:2.

describes the clumsy nature of the "ugly matter" (*massa vetus*), speaks of its "sordid complexion" (*vultus sordes*) and its "miserable deformation" (*vultum miserata deformem*), but also of its "quest for a better look and a more graceful appearance" (*formae melioris opem vultusque decorem quaereret atque suum lugeret silva tumultum*).[30] It is this characterization of matter that Johannes Kepler, the originator of our modern concept of inertial mass, had in mind when he wrote: "All corporeal stuff or matter in the whole world has this virtue, or rather vice (*Unart*), that it is plump and clumsy to move itself from one place to another."[31]

[30] Alanus ab Insulis (of Ryssel), *Anticlaudian* (Venice, 1582), 492 B, 534 D, 442 B, 492 C.

[31] Johannes Kepler, "Annotations to Aristotle's De motu terrae," ex manuscriptis Pulkoviensibus, *Opera omnia,* ed. C. Frisch (Frankfort and Erlangen), vol. 7 (1868), p. 746.

CHAPTER 4

"QUANTITAS MATERIAE" IN MEDIEVAL THOUGHT

The formation of the concept of *quantitas materiae* in the thirteenth century is intimately connected with certain scholastic modifications of Aristotle's theory of matter, which, as we have seen earlier, was a rather problematic and unsettled issue within the confines of the original Aristotelian writings.

Being, in Aristotelian and scholastic thought, was generally divided into that which exists in itself and that which exists in another. The latter category of being was called "accident." Now, an accident may exist through the other or it may be the cause for the other. In the second case it was called "form." The process of the reciprocal transformation of the elements, for example, when water becomes air or air turns to water, was in Aristotelian philosophy an interchange of opposites (the elements); but opposites cannot simply be exchanged; one opposite can be reinstated only where the other has been destroyed. Destruction itself, or the thing destroyed, cannot give rise to another thing (the new opposite). Consequently, the existence of a certain substratum must be postulated in which this exchange can take place.

This substratum, constituting corporeal objects or "bodies," was not regarded as being in itself completely formless. It is already more than the Plotinistic formless matter, the abjected object of Neoplatonic and Judeo-Christian cosmological speculations. It is a combination of prime matter and form. This specific form was generally called "corporeal form" (*forma corporalis*).[1] Thus elementary matter, the common substratum of the four elements, was prime matter and corporeal form. According to Aristotle, prime matter itself was unextended whereas elementary matter, the substance of the elements, naturally had to be regarded as extended. The question then arises: What is the relation between "corporeal form" and "extension"?

Various answers were offered by medieval thought. Avicenna (Ibn Sina)[2] identifies corporeal form with the predisposition of prime matter to assume spatial extension or three-dimensionality. According to Algazel (Al-Ghazali), corporeal form is the cohesiveness or massiveness[3] of matter which constitutes only the basis for the three-dimensionality of matter.[4] For Averroes (Ibn Rushd) it is indeterminate three-dimensionality — extension in three-dimensional space as such — but not the variable and measurable three-dimensionality, which quantity he usually refers to as "determinate three-dimensionality." Determinate dimensionality is an accident, capable of being increased or decreased; indeterminate dimensionality is a form, essential to matter.

Averroes emphasizes the importance of this distinction in his treatise *Concerning the substance of the celestial sphere*.[5] In his examination of the matter and form of celestial bodies Averroes accepts the Aristotelian dictum that the individual difference in

[1] See Chapter 2, p. 22.

[2] *Die Metaphysik Avicennas,* ed. and trans. M. Horten (Halle and New York, 1907), part 2, "Die Substanz," chap. 2, p. 100. Avicenna's identification of corporeal form with susceptibility to three-dimensionality was accepted by Ibn Tufail, Hawarizmi, Gorgani, and Faruqi.

[3] "Massiveness," a term used by H. A. Wolfson in connection with Algazel's concept of corporeal form, is an excellent translation of the original *itsal,* if the meaning of *massa* (= paste) is recalled. See H. A. Wolfson, *Crescas' critique of Aristotle* (Harvard University Press, Cambridge, 1929), p. 101.

[4] Al-Ghazali, *Maqasid al falacifa,* ed. G. Beer (Leiden, 1888).

[5] Averroes, *Sermo de substantia orbis,* completed about 1178. The original Arabic text, *Maqalah fi al-jirm al-samawiy,* is lost. The Hebrew version, *Ma'amar be'ezem ha-galgal,* has been edited recently by Arthur Hyman (thesis, Harvard University, 1953).

existents is brought about by the form in matter,[6] but adds that the very existence of different objects of the same substantial form implies the divisibility of matter. Prime matter, therefore, independent of substantial form, must be endowed with divisibility or quantity.[7]

Narboni, in a commentary to Averroes' treatise, restates Averroes' refutation of Avicenna's and Algazel's theses as follows:

> Avicenna thinks that "body" is a term applying to the substantiality which has the possibility that the three indeterminate dimensions . . . rest in it. This is what is meant by "corporeity" which is the first form existing in matter as yet undistinguished by any other form. This corporeity is not of the nature of dimension which is an accident in the category of quantity and which may change, increase and decrease, as e.g. in the case of three-dimensional wax which changes in respect to roundness and in the case of air which decreases [in quantity]. All bodies have the corporeal form in common and in virtue of it each one is said to be a body. And they differ in virtue of the specific forms through which they are called a particular body. And the corporeal form is not identical with cohesion, for a body can be divided and still remain a body . . . From this it is clear that Avicenna assumes that the corporeal form is other than the dimension, and it is not cohesion as Al Ghazali . . . thought, and cohesion is not essentially necessary to its nature. But corporeal form is other than this, for corporeal form is something which prime matter does not strip off, while the dimensions change, increase and decrease . . .
>
> But Averroes argues against this, maintaining that the "dimensions" are the corporeal form and that the prime matter does not strip them off, but it only puts off their boundary and their limit, and these are the determinate dimensions. For dimensions increase and decrease but they do not change.[8]

This controversy concerning the nature of the corporeal form is of importance for our subject for several reasons. In the first place, it is an expression of a general tendency to find some-

[6] Cf. Aristotle, *Parts of animals,* book 1, chap. 3, 643 a 24.

[7] "Quoniam si non haberet dimensionem, non reciperet insimul formas diversas numero, nec formas diversas specie, sed in eodem tempore non inveniretur nisi una forma." Reference 5.

[8] Moses ben Joshua of Narbonne, *Commentary to the Hebrew text of "De substantia orbis",* quoted from A. Hyman, reference 5, p. 218.

thing that characterizes the essence of matter and yet is different from spatial extension. Secondly, the Averroistic concept of indeterminate dimensions, though slightly modified, becomes an important factor in Aegidius' definition of *quantitas materiae,* the first explicit definition of a concept of mass. Finally, Averroes, in his refutation of Avicenna's concept of corporeal form, argues that substance, according to Avicenna, would already possess an actual form and thus be an actual being (without the addition of further forms). This conclusion, however, would contradict the Aristotelian thesis according to which the elements do not generate from an actual being. Averroes now continues that, had Avicenna interpreted corporeal form as merely the capacity of prime matter for natural motion and as merely the tendency to move to its natural place, then Avicenna's theory would have gained consistency and may become acceptable.[9]

This statement, although made by Averroes more or less in an incidental manner, is most remarkable. For it alludes to the possibility of conceiving the essence of matter in its dynamic behavior. Historically viewed, it is the earliest, although as yet highly inarticulate, expression of a dynamical concept of mass. For the immediate future, however, Averroes' concept of "indeterminate dimensions" was of greater importance.

The new impetus came from theological speculation. As we have shown in the historical analysis of other basic notions in physics, theological reasoning and scholastic cogitation had a decisive impact upon scientific concept-formation.[10] The problem of faith, revelation, and reason, and in particular the question of reconciling eschatological and thaumaturgical tradition with rational thought, were important factors in this respect.

For the concepts of matter and mass, the following three theological topics were of importance: creation, death, and

[9] *Die Epitome der Metaphysik,* trans. S. Van Den Bergh (Leiden, 1924), pp. 63–67.

[10] Cf. Max Jammer, *Concepts of space* (Harvard University Press, Cambridge, 1954), pp. 25–50.

transubstantiation. They corresponded to the problems in natural philosophy of the production, annihilation, and transmutation of matter and consequently were also related to the, as yet primarily metaphysical, principle of conservation of matter.

"Matter cannot be generated nor can it be annihilated, since all that is generated, is generated from matter, and all that perishes, perishes into matter"[11] is a scholastic version of the principle of conservation of matter. The question had to be faced how to reconcile this principle with the dogma of the divine creation of the world. The problem was solved by restricting the applicability of the principle to the framework of the created world itself: "Something created cannot create matter."[12]

Another, perhaps more complicated, example of this theologico-scientific interconnection was a problem raised by a passage in Genesis 2:21, 22: "And the Lord God caused a deep sleep to fall upon Adam, and he slept; and he took one of his ribs, and closed up the flesh instead thereof; And the rib, which the Lord God had taken from man, made he a woman, and brought her unto the man." The creation of Eve, in relation to the principle of conservation of matter, raises a problem for natural philosophy. William of Conches, in his *De philosophia mundi*,[13] already dealt with it and Aegidius Romanus discussed the issue in a query: "Whether Eve could be made out of Adam's rib without the addition of matter?"[14]

With respect to the second theological theme, that of death, the principle of conservation of matter seemed often to have

[11] "Materia non est generabilis nec corruptibilis, quia omne quod generatur, generatur ex materia, et quod corrumpitur, corrumpitur in materiam." *De natura materiae,* chap. 1, art. 6, in *Textus philosophici Friburgenses,* No. 3, ed. Joseph M. Wyss (Louvain and Fribourg, 1953). In contrast to Wyss, M. Grabmann ascribes this text to St. Thomas.

[12] "Creatura igitur materiam creare non potest." *Ibid.,* art. 13.

[13] William of Conches, *Dragmaticon philosophiae* (*Dialogus de substantiis physicis*) (Strassburg, 1567).

[14] "Utrum de costa Adae sine additione materiae potuerit fieri Eva?" *Quodlibet,* book 2, question 11.

supplied a rational justification for the belief in the resurrection of the flesh. Tatian, a Christian apologist of the second century, proclaimed that, although his body may be burned, the universe still retains the matter of his body in the form of ashes.[15] Tertullian, in his treatise *On the resurrection of the flesh,*[16] written at the beginning of the third century A.D., deduces from the imperishability of matter the possibility of a restoration of the flesh after death.

The question of how far the third theological theme, the dogma of transubstantiation, is intrinsically related to the notion of resurrection — Ignatius called the Eucharistic Bread "the medicine of immortality and the antidote against death[17] — does not particularly concern us in the context of our present discussion. What is of importance here is to show how the idea of *quantitas materiae* originated from a conceptual analysis relating to the transubstantiation of the Eucharistic Bread.

In the early ages of the Church the notion of the Eucharist was vague and undefined. In the ninth century the first systematic treatise on the Eucharist, *On the Body and the Blood of the Lord,*[18] was written by the Frankish monk Paschasius Radbertus. Paschasius claimed that the bread and wine are changed into the body and blood of Christ by consecration, the change not being apparent to the senses. From that time until its full adoption by the fourth Lateran Council in 1215 the philosophico-theological meaning of the Eucharist was gradually clarified and the notions involved were more accurately defined. The problem that scholastic theology had to face was how to reconcile the Aristotelian doctrine of substance and accidents with the Chris-

[15] Tatian, *Address to the Greeks* (*Oratio ad Graecos*), ed. E. Schwartz (Leipzig, 1888), 6, 6; see Migne, Series Graeca, vol. 6.

[16] Quintus Septimus Florens Tertullianus, *Liber de resurrectione carnis,* chap. 11; Migne, Series Latina, vol. 2, p. 809.

[17] Ignatius, *Epistles to the Ephesians,* 20. "Pharmacum immortalitatis, antidotum non moriendi." See Migne, Patrologia Graeca, vol. 5, p. 755.

[18] Paschasius Radbertus, *De corpore et sanguine domini* (about 831); see La Bigne, *Bibliotheca veterum patrum,* vol. 6 (Paris, 1609); Migne, vol. 120, pp. 1255–1350.

tian dogma of transubstantiation,[19] according to which the whole substance of the bread changes into the Body of Christ and the whole substance of the wine into the Blood while the accidents ("species") of bread and wine alone remain. Peter Lombard[20] does not yet raise the subtle question concerning the ontological status of the substance and accidents in the Eucharist. Alexander of Hales, the first theologian who incorporated not only the logic but also the physics of Aristotle in his teachings, rejects the idea of consubstantiation (that is, *assumption:* Christ "assumes" the bread and the wine into His Body) or of annihilation of the substance and speaks of a transition;[21] the persistence of the accidents is for him a miracle. How otherwise could these accidents, like weight, density, color, odor of the host, which qua accidents have no independent existence, continue to exist? A rational answer to this question seemed to Alexander impossible.

Thomas Aquinas, however, attempts to advance a solution of this problem. In the third Distinction of the *Commentary to Petrus Lombardus's Book of sentences*[22] Thomas uses for the first time and with reservation the Averroistic notion of indeterminate dimensions. In later passages of the same commentary[23] Aquinas seems to accept fully the Averroistic doctrine of determinate and indeterminate dimensions and finally he makes use

[19] The term "transubstantiation" appeared probably for the first time in the *Exposition of the Canon of the Mass,* attributed to Peter Damiani, about 1072.

[20] Petrus Lombardus, *Sententiarum libri quatuor,* book 4, distinction 11 ("Quomodo fiat ista conversio"); see Migne, Series Latina, vol. 192, p. 1096. Referring to Augustine, Peter says: "Mysterium fidei salubriter credi potest, investigari salubriter non potest."

[21] Alexander Halesius, *Summa universae theologiae* (Venice, 1575), book 4, question 38.

[22] *Commentaria in IV librum sententiarum magist. Petri Lombardi,* distinctio 3, questio 1, articulus 4, in *Sancti Thomae Aquinatis Opera Omnia,* vol. 6 (Parma, 1857), p. 415.

[23] "Non enim una pars materiae diversas formas oppositas et disparatas simul recipere potest. Sed impossibile est in materia intelligere diversas partes, nisi praeintelligatur in materia quantitas dimensiva ad minus interminata." *Ibid.*

Distinctio 12, qu. 1, art. 1: "Utrum accidentia sine subjecto esse Deus facere possit." *Ibid.,* pp. 653–655.

of them rather unreservedly in his explanation of resurrection.[24]

In his *Exposition of Boethius' Book on the Trinity*[25] Aquinas refers explicitly to the notion of indeterminate dimensions as the basis for the principle of individualization. Finally, in his magnum opus, the *Summa theologica,* he employs the Averroistic conception in order to solve the problem of transubstantiation. In part 3, question 77, article 1 of the *Summa* he first proves that the accidents in the Eucharistic sacrament remain without subject[26] and in article 2 he raises the critical question, "Whether in this sacrament the dimensional quantity of the bread and of the wine is the subject of the other accidents?"[27] His answer is: "With the exception of the dimensional quantity all accidents that remain in the sacrament are not in a substance but in the dimensional quantity of the bread and the wine as in a subject."[28]

Aquinas's explanation of the problem of transubstantiation led to a certain looseness and ambiguity in the use of the term "accident" and also to a blurring of the Averroistic distinction between indeterminate and determinate dimensions. It is no longer the actual inherence in a subject — the accidents of the consecrated *hostia* do not inhere in substance — that characterizes accidents, but their aptitude to inhere. This shift in definition contributes to a gradually increasing recognition of an inde-

[24] Distinctio 44, qu. 1, art. 1: "Illud quod intelligitur in materia ante formam, remanet in materia post corruptionem: quia remoto posteriori, remanere adhuc potest prius. Oportet antem, ut Commentator dicit, in 1 Physic. et in lib. de substantia orbis in materia generabilium et corruptibilium ante formam substantialem intelligere dimensiones non terminatas." *Ibid.,* p. 1075.

[25] *Sancti Thomae de Aquino Expositio super Librum Boethii de Trinitate,* ed. B. Becker (Leiden, 1955), p. 143, qu. 4, art. 2: "Dimensiones . . . possunt dupliciter considerari. Uno modo secundum earum terminationem . . . alio modo possunt considerari sine ista determinatione . . . et ex his dimensionibus interminatis materia efficitur materia signata et sic individuat formam."

[26] *S. Thomae Aquinatis Summa Theologica,* vol. 5 (Taurin, 1886), p. 139. "Utrum accidentia remaneant sine subjecto in hoc sacramento."

[27] "Utrum in hoc sacramento quantitas dimensiva panis vel vini sit aliorum accidentium subjectum." *Ibid.*

[28] "Omnia accidentia praeter quantitatem dimensivam, quae remanent in sacramento, quamvis in nulla sint substantia, sunt tamen in quantitate dimensiva panis et vini tanquam in subjecto." *Ibid.*

pendent reality of accidents — a process which in the development of nominalism led to the important result that the concept of an accident without subject in which it inheres does not involve any contradiction.[29]

With these remarks in mind we are now in a position to understand how Aegidius Romanus, a disciple of Aquinas,[30] formed the concept of *quantitas materiae* as a measure of mass or matter, independent of determinations of volume or weight.[31]

Aegidius's point of departure, as expounded in Proposition 44 of his *Theorems concerning the Body of Christ*,[32] is the Thomistic problem concerning the persistence of accidents in the Eucharistic host, in particular those of condensation and rarefaction. Aquinas' solution of taking quantity (*quantitas dimensiva*) as the subject for these accidents is for Aegidius not completely satisfactory, since in the case of condensation (that is, the perceptible increase in density), for instance, what changes is just quantity itself, and quantity qua changing quantity cannot be an accident of quantity qua quantity as subjectum that sustains

[29] Cf. for example, Nicholas of Autrecourt's theory of accidents. J. Maréchal, *Point de départ de la métaphysique* (Paris, 1944), vol. 1, p. 172; also T. Lappe, *Beiträge zur Geschichte der Philosophie des Mittelalters*, vol. 6, part 2 (Münster, 1908), pp. 29–30. The dependence of accidents on a substratum was also disputed by the motazilites Abu'l-Hudhail and al-Juba'i al-Sahili [see *Maqasid al-falasifa*, ed. Ritter (Istanbul, 1929–1930), pp. 310–312 (*Maqualat al-Islamiyyin wa'khtilaf al-Musallin*)] and by al-Yuwaymi in his *Al-Irshad*, ed. Luciani (Paris, 1938), p. 13. A similar process of attaching reality to accidents independent of a substratum can be recognized in modern times in Einstein's rejection of the ether which led to the characterization of "empty space" by at least three numerical "accidents," a velocity of 3×10^{10} cm/sec, a characteristic resistance of 377 ohms, and an average curvature not yet very well measured.

[30] Aegidius Romanus, also called Egidio Colonna or Giles of Rome, was, according to Godfrey of Fontaines, "the greatest teacher at the University of Paris." Cf. N. Mattioli, *Studio critico sopra Egidio Romano*, Antologia Agostiana (Rome, 1896), vol. 1.

[31] For a list of Aegidius's writings see G. Boffito, *Saggio di bibliographica Aegidiana* (Florence, 1911), and for Aegidius's conception of *quantitas materiae*, Anneliese Maier, *Die Vorläufer Galileis im 14. Jahrhundert* (Storia e Letteratura, Rome, 1949).

[32] Aegidius Romanus, *Theoremata de Corpore Christi* (completed 1276); cf. E. Hocedez, *Richard de Middleton* (Louvain, 1925), p. 460.

the change. The difficulty disappears as soon as it is assumed that it is not one and the same quantity that is under discussion. In other words, if we assume that there are two different kinds of quantities, then one kind can sustain (as subject) the variation of the other kind (as accident) and no logical inconsistency is involved.

This theory of *duplex quantitas* thus explains condensation — even in the absence of substance — as a ratio between the two quantities that are called determinate and indeterminate dimensions, the former corresponding to volume and the latter to what was later called *quantitas materiae.*

> It should be understood that in the matter of the bread and the wine as well as in all earthly matter there are two quantities and two kinds of dimensions: determinate and indeterminate dimensions. For matter is so and so much and occupies such and such a volume . . . If it can be shown that it is not the same quantity by which matter is so and so much and by which it has such and such a volume, and, on the other hand, if we can state that the quantity in virtue of which matter is so and so much precedes the quantity in virtue of which it occupies such and such a volume and that in the first kind of quantity, as in a subject, the second kind of quantity is anchored, then it is easy[33]

to solve the problem of transubstantiation. Aegidius thus accepts the Averroistic terminology of determinate and indeterminate dimensions, but their meaning is slightly different: *dimensiones indeterminatae* is now the name for quantity of matter (*quantitas*

[33] "Notandum ergo, quod in materia panis et vini et in materia omnium generabilium et corruptibilium est duplex quantitas et duplex genus dimensionum. Sunt enim ibi dimensiones determinatae et indeterminatae: est enim materia generabilium et corruptibilium tanta et tanta et occupat tantum et tantum locum . . . Si ergo ostendere poterimus quod non est eadem quantitas per quam materia est tanta et per quam occupat tantum locum, et rursum, si poterimus declarare quod quantitas illa, per quam res est materia tanta, praecedit quantitatem illam, per quam materia occupat tantum locum, et quod in prima quantitate . . . tamquam in subiecto fundatur alia quantitas . . . facile erit . . ." Proposition 44, *Theoremata de Corpore Christi* (42 leaves; Balthasar de Hyrubia (Rubbiera), Bologna, 1481), Boston Medical Library, Boston, Mass. Aegidius restated and expounded these arguments in his *Commentaries to the eight books of Aristotle's Physics* (1277) (*Commentaria in octo libros phisicorum Aristotelis*), book 4, chap. 9, and in his *Metaphysical questions* (*Metaphysicales quaestiones*), book 8, question 5.

materiae) and *dimensiones determinatae* the name for volume, the measurable, determinable spatial quantification. This distinction, argues Aegidius, follows from the fact that the variation of one of them does not imply a variation of the other, as, for instance, in the process of rarefaction in which the determinate dimensions increase while the quantity of matter remains invariant. Their essential heterogeneity is also evident, he claims, from the fact that an external agent may affect one without affecting the other, as, for instance, in the above-mentioned example of a (volume-) expanding force. Indeed, in conformity to the principle of conservation of matter, no natural force exists that could affect the indeterminate dimensions.

Aegidius's insistence on the ontological priority of indeterminate dimensions ("quantitas illa praecedit quantitatem illam") was for him a relation necessary for the explanation of the condensation or rarefaction of the Eucharistic Bread, since the indetermined serve, in his view, as the subject for the determined dimensions, or, in a more modern terminology, mass (as quantity of matter) is the carrier of spatial extension. It is interesting to note that, in contrast to Cartesian physics, the classical mechanics of the seventeenth century fully adopted this priority, although on different grounds. The fundamental concept in Newtonian physics is the mass-particle (without spatial extension) and not an elementary volume without mass!

Aegidius thus postulated the existence of a new quantitative measure of matter that is different from volume determination. Needless to say, the concept of weight in Aegidius's natural philosophy, still conceived along Aristotelian lines, cannot play the role of such a measure. The Aegidian innovation is, indeed, a radically new conception, a new "dimension" in the modern technical sense of the word. Anneliese Maier calls it "undoubtedly one of the most modern ideas in the natural philosophy of the Schola."[34] And yet, in spite of Aegidius's clear presentation of his ideas, his innovation gained little recognition among the

[34] Reference 31, p. 46.

Schoolmen. In fact, Aegidius himself, after 1289, renounced his ideas concerning the new concept of *quantitas materiae*. In his *Quodlibet* 4, question 1,[35] he emphatically revokes his former statements and redefines indeterminate dimensions, again in conformity with Averroes, as that in virtue of which matter occupies space, and determinate dimensions as that in virtue of which matter occupies a determinate space or volume. Nor do his contemporaries or successors accept his innovation. If mentioned at all, it is generally rejected. Thus Thomas of Sutton, a member of the English Thomistic school, at the end of the thirteenth century criticized Aegidius's theory in a *Quodlibet* and proclaimed the essential identity of the two quantities.[36] Godfrey of Fontaines, whose *Quodlibeta* enjoyed great popularity throughout the Middle Ages, refers in his *Quodlibet* XI to Aegidius's (unrevised) solution of the Eucharistic problem but eventually accepts the existence of only one quantity in matter which depends on the density of the substance.

Thus a concept which, owing to the inherent lack of operational significance would in any case have proved itself unacceptable for modern science, enjoyed only a very short existence. Aegidius's innovation was, however, the first explicit definition of a new quantitative aspect of matter and as such is of importance for a historical study of the concept of mass.

[35] *Aegidii Columnae Romani quodlibeta,* ed. De Coninck (Louvain, 1646).
[36] Quodlibet II, question 18.

CHAPTER 5

THE CONCEPTUALIZATION OF INERTIAL MASS

The concept of *quantitas materiae,* although seldom considered worthy of being distinctly defined, continued nevertheless to play a not inconsiderable role in medieval science. The theory of impetus, for example, as expounded by John Buridan and his school in the fourteenth century, required the concept of quantity of matter in order to account for the experimental fact that the action of the same motive force causes a stone to move farther than a feather or a piece of iron farther than a piece of wood of the same size. The explanation put forward by Buridan admits that

> the cause of this is that the reception of all forms and natural dispositions is in matter and by reason of matter. Hence by the amount more there is of matter (*quanto plus de materia*), by that amount can the body receive more of that impetus and more intensely (*intensius*). Now in a dense and heavy body, other things being equal, there is more of prime matter than in a rare and light one.[1]

Buridan's idea of a proportionality between impetus and quantity of matter conforms to the general rule of a quantitative correspondence between form and matter.[2] This rule is considered

[1] Buridan, *Quaestiones super octo libros Physicorum,* book 8, question 12 (Paris, 1509); English translation of question 12 in Marshall Clagett, *The science of mechanics in the Middle Ages* (University of Wisconsin Press, Madison, 1959), document 8.2, p. 535.

[2] "Forma multiplicatur et dividitur secundum multiplicationem et divisionem materiae in qua est." Buridan refers to this proportionality repeatedly in book 7 of his commentary on physics, questions 7 and 8; see reference 1.

to apply not only to substantial forms but also to forces and qualities in general, as impetus, heat, coldness, dryness. The greater capacity for impetus in a big rotating mill wheel, contends Buridan, is the reason why it is more difficult to stop its motion than it is in the case of a smaller wheel, *ceteris paribus*. The quantity of matter present in a body determines, according to Buridan and his school, quite generally the resistance that a physical object exerts against the moving force. Although resistance here, as in Aristotle, is still considered a real force, it is surprising to see how closely the impetus theory approached the concept of inertial mass. But impetus was not momentum and resistance not inertia.

Similar examples for the implicit recognition of "quantity of matter" in connection with the impetus theory are found in the *Questions on the eight books of the Physics of Aristotle* by Albert of Saxony[3] and in *The treatise on the heaven and the world* by Nicole Oresme.[4]

Another subject that naturally suggests the use of our concept is the problem of rarefaction and density, which — for itself and apart from its implications for theology — continued to engage the medieval mind. Thus Richard Swineshead dedicated considerable space in his *Book of calculations* to the subject of rarefaction and condensation. "A thing is rarefied in the ratio of its quantity to its matter and increases in density according to the ratio of its matter to its quantity."[5] In what may be considered an early anticipation or analogy of the equation of continuity in classical field theories, the Calculator states:

If two bodies are of unequal size and unequal density, and the ratio of the quantity of the denser body to the quantity of the other is less than

[3] Albertus de Saxonia, *Quaestiones super octo libros Physicorum* (Venice, 1500?).

[4] Nicole Oresme, *Le livre du ciel et du monde,* ed. A. D. Menut and A. J. Denomy, *Mediaeval studies,* vols. 3–5 (Toronto, 1941–43).

[5] Richard Swineshead, *Liber calculationum* (Padua, 1477), fol. 17 r. Quoted from Lynn Thorndike, *A history of magic and experimental science,* vol. 3 (Columbia University Press, New York, 1934), p. 378.

the ratio of their respective densities, if they gain or lose density at equal speed, the denser body will lose or acquire quantity more slowly than the rarer body.[6]

Lynn Thorndike, in an article on "An anonymous treatise in six books on metaphysics and natural philosophy"[7] of the fourteenth century remarks that "something approaching the conception of mass seems involved in the statement that many ancients and moderns distinguish *quantitas continua seu molis* from *res quanta*."

These examples show that already before the rise of classical mechanics in the sixteenth and seventeenth centuries the concept of quantity of matter was needed for the formulation of physical laws. Although this need was strongly felt, the notion was only a most vague and nebulous concept. Even the Italian natural philosophy of the Renaissance, which on the whole made an important contribution to the concept formation of modern science, did surprisingly little to clarify the situation. Galileo, for instance, in his *Assayer,* the exposition of his philosophy of science, remarks:

> Now I say that whenever I conceive any material or corporeal substance, I immediately feel the need to think of it as bounded, and as having this or that shape; as being large or small in relation to other things, and in some specific place at any given time; as being in motion or at rest; as touching or not touching some other body; and as being one in number, or few, or many. From these conditions I cannot separate such a substance by any stretch of my imagination.[8]

Galileo here enumerates the primary qualities of matter: shape, size, location, contiguity, number, motion. All these qualities are of a geometric (shape, size, location, contiguity), arithmetic (number), or kinematic (motion) character. Nowhere in this list does he mention any non-geometrico-temporal aspect of matter. Only Evangelista Torricelli — long after Kepler's explication of the concept of mass — added "mass" as a "dimension"

[6] *Ibid.,* fol. 21 v.

[7] *Philosophical Review 11,* 317–340 (1931).

[8] *Discoveries and opinions of Galileo,* trans. Stillman Drake (Doubleday, New York, 1957), p. 274.

among these primary qualities of matter. Giorgio de Santillana, in an interesting footnote to his edition of Galileo's *Dialogue on the great world systems,*[9] describes Galileo's view on mass as follows:

> Thus, there are mathematical properties, inherent in matter; but mass, although mathematizable, is not one of them, for it is another name for matter itself and distinguishes it from abstract matter which is geometry. Physical reality and mass are two names for the same thing which possesses motion, whereas geometrical shapes do not possess it. Hence mass cannot be defined in terms of anything else; it is a *primum.*

There are, however, a few passages in Galileo's writings that suggest the idea of inertial mass. In the *Dialogue on the great world systems,* for instance, Salviati asks the question "whether there is not in the body, besides the natural inclination to the contrary term (as in heavy bodies that have a propensity to the motion downwards the resistance is to the motion upwards), another intrinsic and natural quality which makes it averse to motion."[10] Galileo could not yet realize that it is exactly this "intrinsic quality" that makes the velocity of free fall independent of the weight of the falling object, a result for the verification of which he performed his historic experiments with metallic spheres on inclined planes.

Galilean physics did not work out a distinct formulation nor did it even establish an explicit recognition of what proved to be, apart from length and time, the most fundamental concept in classical physics. It was Keplerian astronomy that filled the lacuna and thus completed the foundation on which Newton's genius could build the majestic structure of classical mechanics.

In possession of Tycho Brahe's accurate observational material, Kepler realized that the traditional notion of a circular and uniform motion of celestial bodies had to be abandoned. In 1609 Kepler discovered that the assumption of elliptical orbits for planetary motion agrees perfectly well with the ob-

[9] Galileo Galilei, *Dialogue on the great world systems,* ed. Giorgio de Santillana (University of Chicago Press, Chicago, 1953), p. 252.

[10] *Ibid.,* p. 228.

servational data at his disposal. The replacement of circles by ellipses, however, raised a new difficulty. Circular motion, since the days of Plato, had been the natural motion of the planets because of the simplicity, perfection, and continuity of the circle.[11] The problem that Kepler consequently had to face was this: do planets in elliptical paths still move "naturally"? Is the concept "naturalness" really a primary notion or is it reducible to a causal relation based on more fundamental natural laws? The search for a dynamical explanation of planetary motion thus became Kepler's preoccupation.

In our historical analysis of the concept of force[12] we have shown how Kepler, in search of such a dynamical explanation, developed the idea of physical force from the rudimentary notion of an *anima motrix* and through the concept of a *species immateriata,* and how astronomical considerations concerning the dependence of planetary velocity upon distance were instrumental for this conceptual process. Kepler's conceptualization of mass follows the same pattern as his conceptualization of force. In fact, they are two complementary aspects of the same intellectual process, just as "mass" is the complementary notion to "force." Kepler's concept of force developed from the idea of motory intelligences, souls, or pure forms, his concept of mass from that of matter. The traditional metaphysical antithesis of *forma* and *materia* is the common background for the two concepts. A factor which, as Kepler discovered, works in opposition to motory forces must necessarily belong to the realm of matter, since it is the nature of matter, according to Neoplatonic tradition, to constitute an impediment to the realization of form.

An early allusion to such a resistance of matter to motion, and to the tendency for rest inherent in matter, is found in

[11] See, for instance, Plato, *Timaeus,* 40; Aristotle, *De caelo,* 268 b 15–21, 269 a 19–20, 24–26, 270 b 32–33; Aristotle, *Physics,* 223 b 21–24, 241 b 20, 265 a 15; Simplicius, *Commentarium in Aristotelis De caelo,* 498 a 46–b 3.

[12] Max Jammer, *Concepts of force* (Harvard University Press, Cambridge, 1957), pp. 81–93.

Kepler's *De stella nova in Pede Serpentarii.*[13] Kepler does not yet use the term "inertia," but explains the resistance (*renitentia*) by an analogy to weight (*pondus*). A few years later he distinguishes one from the other with more discrimination.[14] In his *New astronomy with commentaries on the motions of Mars*[15] Kepler infers the material nature (*natura materiata*) of the planets from their apparent and inherent propensity for absence of motion. In the same *Commentaries* Kepler uses the concept of mass, which he calls *moles,* in a dynamical sense as opposing mutual attraction:

> If two stones were placed in any part of the world, near each other yet beyond the sphere of influence (*orbem virtutis*) of a third related body, the two stones, like two magnetic bodies, would come together at some intermediate place, each approaching the other through a distance in proportion to the mass of the other.[16]

It is interesting to note that this relation could have served in principle as an operational definition of mass ratios somehow analogous to Mach's more recent definition of (relative) mass. But, as we have mentioned in connection with a similar situation regarding Kepler's conceptualization of force,[17] such methodo-

[13] Kepler, *De immensitate sphaerae fixarum in hypothesibus Copernici: deque Novi Sideris magnitude, Opera omnia,* ed. C. Frisch (Frankfort and Erlangen), vol. 2 (1859), "Nam quietem quidem loci seu ambientis corporis affectant renitentia et quodam quasi pondere (quid ridetis coelestium inexperti philosophastri, rerum imaginarium copia locupletes, verarum egentissimi?), ex quo singulis suae obveniunt periodi temporum (nam quod motorem attinet, is unicus est et uniformiter movet)."

[14] *Epitome astronomiae Copernicanae* (1618), book 4, part 3, *Opera omnia,* vol. 6 (1896), p. 374: "Pondus ergo tribuis planetae? Dictum est in superioribus pro pondere considerandum esse naturalem illam materialem renitentiam seu inertiam . . ."

[15] *Astronomia nova aitiologetos seu physica coelestis, tradita commentariis de motibus Stellae Martis (1609),* part 3, chap. 34, *Opera omnia,* vol. 3 (1860), p. 305: "Necesse est igitur, ut planetariorum globorum natura sit materiata, ex adhaerente proprietate inde a rerum principio prona ad quietem seu ad privationem motus."

[16] *Ibid.,* p. 151: "Si duo lapides in aliquo loco mundi collocarentur propinqui invicem extra orbem virtutis tertii cognati corporis, illi lapides ad similitudinem duorum magneticorum corporum coirent loco intermedio, quilibet accedens ad alterum tanto intervallo, quanta est alterius moles in comparatione."

[17] See reference 12, pp. 88–89.

logic considerations were still beyond Kepler's conceptual scheme. In fact, nowhere does Kepler refer to the implications of such statements as expressed in his *Astronomia nova.* The important property of matter is for him merely its inherent tendency to remain at its place. In the *Tertius interveniens* (1610) he states emphatically: "I, for my part, say that the celestial spheres have the property of abiding at any place in the heavens wherever they are found, unless they are drawn along."[18]

In the *Epitome astronomiae Copernicanae* (1618), the earliest textbook of Copernican astronomy, the influence of Neoplatonic thought upon Kepler's reasoning is evident: "Every celestial sphere, because of its materiality (*ratione suae materiae*) has a natural inability to move from place to place, a natural inertia or rest whereby it remains in every place where it is set by itself."[19]

The transformation of metaphysical speculation into physical reasoning, a process that makes Kepler's contribution to the formation of modern scientific concepts so important, is carried out in the following conclusion:

> If the matter of celestial bodies were not endowed with inertia, something similar to weight, no force would be needed for their movement from their place; the smallest motive force would suffice to impart to them an infinite velocity. Since, however, the periods of planetary revolutions take up definite times, some longer and others shorter, it is clear that matter must have inertia which accounts for these differences.[20]

[18] *Opera omnia,* vol. 1 (1858), p. 590: "Für mein Person sage ich, dass die Sternkugeln diese Art haben, dass sie an einem jeden Ort dess Himmels, da sie jedesmal angetroffen werden, stillstehen würden, wann sie nicht getrieben werden solten."

[19] *Opera omnia,* vol. 6 (1896), p. 341: "Globus aliquid coelestis habet tamen ratione suae materiae naturalem adinamian transeundi de loco in locum, habet naturalem inertiam seu quietem qua quiescit in omni loco, ubi solitarius collocatur."

[20] *De causis planetarum, ibid.,* p. 342: "Nam si nulla esset inertia in materia globi coelestis, quae sit ei velut quoddam pondus, nulla etiam opus esset virtute ad globum movendum: et posita vel minima virtute ad movendum, iam causa nulla esset, quin globus in momento verteretur. Iam vero cum globorum con-

It is certainly no exaggeration to say that a statement such as this anticipates in a qualitative manner Newton's second law of motion.

A few pages later Kepler describes the process of motion as an encounter of two opposing factors. In planetary motion, he claims, the "transporting power (*potentia vectoria*) of the sun and the impotence of the planet (*impotentia planetae*) or its material inertia strive against each other."[21] Inertia, according to Kepler, is thus not only the inability of matter to transport itself from place to place, but also has, so to say, an active aspect, to be "repugnant" to motion imparted from without. And this resistance or "repugnance" stands in direct proportion to the quantity of matter (*copia materiae*). "Inertia or opposition to motion is a characteristic of matter; it is stronger, the greater the quantity of matter in a given volume."[22]

This statement of Kepler's is of great importance from the historical point of view. For it connects the notion of *quantitas materiae* of the Schoolmen, or, in Kepler's terminology, the *copia materiae*, with his new concept of inertial mass. It is, of course, possible to interpret the expression "quantity of matter confined in a given volume" as denoting "density," in which case Kepler obviously conceives his "inertia" as proportional to the density of matter. This interpretation finds some support in a note that Kepler added to the second edition (1621) of his

versiones fiant in certo tempore, quod in alio planeta est longius, in alio brevius. Hinc apparet, inertiam materiae non esse ad virtutem motricem, ut nihil ad aliquid." This passage appears under the caption: "Unde probas, materiam coelestium corporum reniti suis motoribus et ab iis, velut in libra pondera a facultate sua motrice? Probatur hoc primo ex periodicis temporibus convolutionis globorum singulorum circa suos axes, ut Telluris tempore diurno, Solis tempore 25 dierum circiter."

[21] *Ibid.*, pp. 345–346: "At dictum est hactenus, praeter hanc vim Solis vectoriam esse etiam naturalem inertiam in planetis ipsis ad motum, qua fit, ut inclinati sint, materiae ratione, ad manendum loco suo. Pugnant igitur inter se potentia solis vectoria et impotentia planetae seu inertia materialis."

[22] *Ibid.*, pp. 174–175: "Materiae enim . . . propria est inertia, repugnans motui, eaque tanto fortior, quanto major est copia materiae in angustum coacta spatium."

Prodromus dissertationum cosmographicarum continens mysterium cosmographicum, originally published in 1596:

> Planetary bodies . . . are not to be considered as mathematical points but obviously as material bodies endowed with something like weight or rather an intrinsic faculty of resistance to motion which is determined by the volume of the body and the density of its matter.[23]

Although Kepler did not yet scientifically systematize the new concept, a task that Newton completed a few decades later, he nevertheless conferred on the concept of matter its scientific status as inertial mass. It is perhaps this consideration which made Leibniz contend that it was Kepler who was the first to advance the idea of inertia and that Descartes borrowed the concept from Kepler. One has, however, to admit that with Kepler the concept of inertia refers exclusively to the impossibility of spontaneous motion or the resistance to a transition from rest to motion (acceleration). In the passage already quoted,[24] he speaks explicitly of the "plumpness" of matter and its inability to transfer itself from place to place. The significance of inertia for the continuation of motion once imparted to the body was hardly considered by Kepler. To credit Kepler with being the originator of "inertia," as Leibniz did, may easily lead to misconceptions, if used in connection with Descartes's "inertia," which primarily was the conservation of momentum in motion. Only if rest and uniform motion are regarded as dynamically equivalent and as physically identical states, described from different reference systems, can "Keplerian inertia" be representative of the more general "Cartesian-Newtonian inertia." But Kepler was, of course, still far away from such relativistic considerations. It is the same misunderstanding

[23] "In caput decimum sextum notae auctoris," *Opera omnia,* vol. 1, p. 161: "Corpora planetarum in motu seu translatione sui circa Solem non sunt consideranda ut puncta mathematica, sed plane ut corpora materiata et cum quodam quasi pondere (ut in libro de stella nova scripsi) hoc est, in quantum sunt praedita facultate renitendi motui intrinsecus illato pro mole corporis et densitate materiae."

[24] See Chapter 3, reference 31.

that apparently induced Auguste Comte to call the principle of inertia "la loi de Képler." [25]

In order to appreciate the full significance of Kepler's contribution to the development of the concept of inertial mass it is important to compare Kepler's conceptual scheme with that of the 14th-century schoolmen. In fact, Willam of Alnwick, Oresme, Buridan, Albert of Saxony, and others made extensive use[26] of notions such as *inclinatio ad quietem, inclinatio ad non moveri, inclinatio ad motum oppositum* in their discussions concerning celestial motion. Since, however, in their view the moving forces behind the celestial spheres are still spiritual intelligences, their discussions were not concerned with "natural forces" (*vires naturales*) but with "voluntary forces" (*vires voluntariae*) not subject to physical laws. Kepler, by associating *inertia* with *copia materiae,* made the metaphysical notion of *inactivity* ("plumpness") into what at his time might have been considered a scientific concept.

[25] Auguste Comte, *Cours de philosophie positive* (Paris, 1842), vol. 6, p. 682.

[26] Cf. Anneliese Maier, *Zwischen Philosophie und Mechanik* (Storia e Letteratura, Rome, 1958), pp. 189–236.

CHAPTER 6

THE SYSTEMATIZATION OF MASS

For the history of concept formation it is often useful to distinguish three stages in the development of the concept: (1) conceptualization (that is, the very process of formation); (2) systematization (that is, the incorporation of the newly formed concept into the syntax of the scientific system); and (3) formalization (that is, the formal definition of the concept within the texture of the deductive presentation of the science). These stages are, of course, mutually interrelated and often so much intermingled that a differentiation becomes a hopeless task. Nor do they always appear in the same chronological order.[1]

As far as the concept of mass is concerned, the first stage, its conceptualization, was, as we have seen in the last chapter, essentially the work of Johannes Kepler. The second stage lagged behind for a rather long time so that practically a break occurred in the continuity of the development of our concept. The reason for this delay has to be attributed to the rise of Cartesian physics in the first half of the seventeenth century.

Contrary to Leibniz's statement, not only did Descartes not borrow the (Keplerian) notion of inertia from Kepler, he even explicitly rejected this idea for the conceptual construction of

[1] This is especially evident in modern physics. In Dirac's discovery of the positron, for instance, the formalization preceded the conceptualization.

his system of the physical universe. In a letter to Mersenne, of December 1638, he declares: "I do not recognize any inertia or natural tardiness in bodies."[2] The only place where he qualifies his position is in a letter to de Beaune, of April 30, 1639; in this letter he discusses the case of a body that strikes the earth, which contains, say, a thousand times more matter (*mille fois plus de matière*) than the body itself; owing to the fact that in the course of the impact the transfer of momentum imparts to the earth only a thousandth part of the velocity of the body, one can say that "in this sense, the more matter a body contains, the more it has natural inertia."[3]

His *Principles of philosophy,* the systematic presentation of his scientific theories, completely ignores such an intrinsic property of matter. For the essence of matter, according to Descartes, is spatial extension; geometry and the principle of conservation of momentum (more precisely, of the Cartesian *quantitas motus*) regulate the behavior of physical objects. Extension is the only essential attribute of matter.[4] Thus, even if he discusses the collision of bodies that in our modern terminology would be described as possessing unequal masses, Descartes says: "If body *B* is . . . greater than *C*," or "If the resting body *C* is smaller than *B* . . ."[5]

Quantity of matter, for Descartes, is volume. Weight and

[2] "Je ne reconnais aucune Inertie ou tardiveté naturelle dans les corps." *Oeuvres de Descartes,* ed. Charles Adam and Paul Tannery (Cerf, Paris), vol. 2 (1898), pp. 466–467.

[3] *Ibid.,* pp. 543–544: "Et pour ceque, si deux corps inégaux reçoivent autant de mouvement l'un que l'autre, cette pareille quantité de mouvement ne donne pas tant de vitesse au plus grand qu'au plus petit, on peut dire, en ce sens, que plus un corps contient de matière, plus il a d'Inertie Naturelle."

[4] "Naturam materiae . . . non consistere in eo quod sit res dura, vel ponderosa, vel colorata, vel alio alioquomodo sensus afficiens: sed tantum in eo quod sit res extensa in longum, latum et profundum." *Principia philosophiae,* pars 2, sec. IV; *Oeuvres,* vol. 8 (1905), 42.

[5] *Principia philosophiae,* part 2 (On the principles of material things), trans. Giorgio de Santillana and Walter H. Pitts (M.I.T. Work-sheets, Cambridge, Mass., 1958). The original text has: "si B esset tantillo majus quam C . . ." and "si corpus quiescens C esset minus quam B . . ." *Oeuvres,* vol. 8, pp. 68–69.

gravity, in his theory the result of a rather complicated kinetics, are merely accidental features and in general stand in no proportion to quantity of matter.[6] It is difficult to see how Descartes could have ignored the fact that two geometrically equivalent bodies, like a solid and a hollow sphere of equal radii, move differently when placed in identical relations with the same other bodies.

Descartes's concept of matter as extension also ran counter to contemporary theology. If the nature of matter is extension and if nothing else supports this extension, the Thomistic (as well as the Aegidian) explanation of the Eucharistic transubstantiation loses, so to say, its ground. That such an objection was indeed raised against Descartes is evident from a letter of Gabriel Thibaut to Mersenne: "If quantity is not different from corporeal substance, since after the conversion of the Bread and the Wine one perceives quantity, what, then, is this substance?"[7] An identification of matter with extension makes the Eucharistic dogma void of any real content. In view of this incompatibility of Descartes's concept with the teachings of the Church, the Congregation of the Index condemned his works on November 20, 1663.[8]

Kepler's concept of inertia, the notion of an intrinsic propensity of matter for rest, was rejected not only by Descartes and his school but also, for instance, by Pierre Gassendi, an opponent of Cartesian intellectualism. In his *De motu impresso a motore translato* he rebuts Kepler[9] and tries to show, referring

[6] "Ejus quantitatem non respondere quantitati materiae cujusque corporis. Unde fit, ut ex sola gravitate non facile possit aestimari, quantum in quoque corpore materiae terristris contineatur." *Principia philosophiae,* part 4, section 25; *Oeuvres,* vol. 8, p. 214.

[7] Letter to Mersenne, April 1, 1647, *Oeuvres,* vol. 5, p. 69. Cf. also Arnauld's letter to Descartes (De re quanta a locali extensione non distincta), June 3, 1648, *ibid.,* vol. 5, p. 190.

[8] Cf. G. Daniel, *Le voiage du monde de Descartes* (Paris, ed. 3, 1720), pp. 140–143.

[9] "Non habere eam (i.e. materia) ad motum resistentia, quam vulgo concipimus." *De motu impresso a motore translato,* 1st letter, sec. 15, *Opera omnia* (Florence, 1727), vol. 3.

to the oscillatory motion of the pendulum, that matter has no resistance to motion.

These arguments, however, even before the time of Newton, did not meet with universal approval. The fundamental change, occurring at that time, in the concept of the dynamics of free fall, in particular, strengthened the position of Kepler. The scholastics, in conformity with Aristotle, considered gravity a factor residing in the heavy body itself. Otherwise, their theory of levity would have encountered serious conceptual difficulties. Now, such an attitude, as E. J. Dijksterhuis has pointed out,[10] was not conducive to the idea that matter has a certain passivity of its own. As soon, however, as gravity is regarded as an activity supplied from without, rather than an inherent principle of motion, the situation is different.

The new concept of gravity, as propounded by Giovanni Alphonso Borelli and Isaac Beeckmann, thus gave support to a dynamical concept of mass. Giovanni Battista Baliani, in particular, in his *De motu gravium* (1638) distinguishes between an active principle, extraneous to the gravitating body (*pondus*), and a passive principle inherent in matter (*moles*). The constant proportionality of these two prevents the heavier body from overtaking the lighter one in the process of free fall.

Another approach conducive to the concept of inertial mass was the study of centrifugal forces. Christiaan Huygens in his treatise *On centrifugal force*[11] investigated the magnitude of the centripetal force of a rotating or revolving body. When particles move with equal speed along circles of equal radii, he concludes, the centripetal forces are to each other as their "weights" or

[10] E. J. Dijksterhuis, *Die Mechanisierung des Weltbildes* (Springer, Berlin, Göttingen, Heidelberg, 1956), p. 409.

[11] Christiaan Huygens, *De vi centrifuga* (Opustula postuma; Leiden, 1703). Huygens completed his study of circular motion about 1659 and published his conclusions in his *Horologium oscillatorium* in 1673. The full text of this study appeared, however, only posthumously in 1703. "Unde etiam concludemus vires centrifugas mobilium inaequalium sed in circulis aequalibus aequali velocitate latorum esse inter se sicut mobilium gravitates, seu quantitates solidas." *Oeuvres complêtes,* ed. Société hollandaise des Sciences (Martinus Nijhoff, The Hague), vol. 16 (1929), p. 267.

their "solid quantities." Since the centripetal force, in modern terminology, is given by the formula $F = mv^2/r$, the forces exerted on two bodies moving with the same velocities v on circles with the same radius r obviously satisfy the relation $F_1 : F_2 = m_1 : m_2$. It was apparently this relation that Huygens had in mind when he spoke of "the solid quantities" (*quantitates solidas*) of the bodies under discussion.[12]

More important still in lending support to the final establishment of the new concept of inertial mass was the systematic study of impact phenomena, both of elastic and of inelastic bodies, carried out by Marci, Wallis, Wren, and Huygens. Although the term "mass" does not appear in the early descriptions and investigations of such impact phenomena, it is clear that the idea of mass is often implicitly involved. Take for example Huygens's study of the elastic head-on collision of two "unequal" bodies A and B, as outlined in his *De motu corporum ex percussione* (1668). In Proposition IX of this work Huygens shows how to find the velocities of A and B after collision in terms of their initial velocities. In his derivation, which is based on what in modern terminology is called the principle of conservation of kinetic energy and which seems to have been known to Huygens as early as 1652, obviously the ratio of the inertial masses of A and B has to be taken into consideration. The text of Proposition IX clearly shows that Huygens was aware of this fact; but, not yet having a distinct term for this idea, he could speak only of the "magnitudes" of A and B: "sicut B ad A magnitudine."[13] Finally, the important discovery, made by Jean

[12] Henry Crew, *The rise of modern physics* (Williams & Wilkins, Baltimore, 1928), p. 121, sees in this statement of Huygens's the earliest quantitative concept of mass.

[13] "Propositio IX. Datis corporibus duobus inaequalibus, directe sibi occurrentibus . . . ; invenire celeritates quibus utraque post occursum ferentur . . . Dividatur AB in C ut sit AC ad CB, sicut B ad A magnitudine. . ." *Oeuvres complètes de Christiaan Huygens* (The Hague, 1929), vol. 16, p. 65. For further details on the importance of the early studies of impact phenomena for the formation of fundamental concepts in classical mechanics, see Max Jammer, *Concepts of force* (Harvard University Press, Cambridge, 1957), p. 126.

Richer in 1671, that the weight of a body is a function of the locality, has to be mentioned as an experimental vindication of the correctness of the conceptual basis involved.

Kepler's conception of the inertial behavior of matter, the important conclusions drawn from impact experiments, and the dynamics of rotational motion, all these trends and results now converge in the work of Isaac Newton and lead to a systematization of the concept of mass. Newton, who generally uses the term "quantity of matter" (*quantitas materiae*) or simply "body" (*corpus*) in preference to "mass,"[14] defines the concept in Definition 1 of his *Mathematical Principles of Natural Philosophy:* "The quantity of matter is the measure of the same, arising from its density and bulk conjointly."[15] He adds the following explanation:

> Thus air of a double density, in a double space, is quadruple in quantity; in a triple space, sextuple in quantity. The same thing is to be understood of snow, and fine dust or powders, that are condensed by compression or liquefaction, and of all bodies that are by any causes whatever differently condensed. I have no regard in this place to a medium, if any such there is, that freely pervades the interstices between the parts of bodies. It is this quantity that I mean hereafter everywhere under the name of body or mass (*corporis vel massae*). And the same is known by the weight of each body, for it is proportional to the weight, as I have found by experiments on pendulums, very accurately made, which shall be shown hereafter.

Before we discuss in detail Newton's definition of mass, let us refer to a few more statements in the *Principia* relevant to this discussion. The concept of density, part of the definiens in Definition 1, was not explicitly defined in the first edition of the *Principia*. Only in Book 3, *The system of the world,* Proposition 6, Corollary 4, are "bodies of the same density" defined as those "whose inertias are in the proportion of their bulks."[16] Definition

[14] Cf. V. G. Friedman, "On Newton's theory of mass," *Uspekhi Fizicheskikh Nauk 61,* 451 (1957).

[15] *Sir Isaac Newton's Mathematical principles of natural philosophy and his System of the world,* Motte's translation revised by F. Cajori (University of California Press, Berkeley, 1947), p. 1: "Quantitas materiae est mensura ejusdem orta ex densitate et magnitudine conjunctim."

[16] *Ibid.*, p. 414.

2 introduces "quantity of motion" (*quantitas motus*) or "momentum" as it is called today: "The quantity of motion is the measure of the same, arising from the velocity and quantity of matter conjointly." The following explanation is added: "The motion of the whole (*motus totius*) is the sum of the motion of all the parts; and therefore in a body double in quantity, with equal velocity, the motion is double; with twice the velocity, it is quadruple." Definition 3 describes the *vis insita,* or innate force of matter, as "a power of resisting, by which every body, as much as in it lies, continues in its present state, whether it be of rest, or of moving uniformly forwards in a right line." [17] The explanation, following this definition, is of particular importance:

> This force is always proportional to the body (*suo corpori*) whose force it is and differs nothing from the inactivity of the mass (*inertia massae*), but in our manner of conceiving it. A body, from the inert nature of matter, is not without difficulty put out of its state of rest or motion. Upon which account, this *vis insita* may, by a most significant name, be called inertia (*vis inertiae*) or force of inactivity. But a body only exerts this force when another force, impressed upon it, endeavors to change its condition; and the exercise of this force may be considered as both resistance (*resistentia*) and impulse (*impetus*); it is resistance so far as the body, for maintaining its present state, opposes the force impressed; it is impulse so far as the body, by not easily giving way to the impressed force of another, endeavors to change the state of that other. Resistance is usually ascribed to bodies at rest, and impulse to those in motion; but motion and rest, as commonly conceived, are only relatively distinguished; nor are those bodies always truly at rest, which commonly are taken to be so.[18]

Finally, it is important to note that the three laws of motion, which follow after these definitions, make no explicit use of the concept of mass.[19]

[17] "Materiae vis insita est potentia resistendi, qua corpus unumquodque quantum in se est, perseverat in statu suo vel quiescendi vel movendi uniformiter in directum."

[18] *Ibid.,* p. 2.

[19] The second law uses the notion of motion (*motus*) in which the concept of mass, according to Definition 2, is implicitly contained. The explanation following Definition 8 of the motive quantity of centripetal force is the only passage in the early parts of the *Principia* where the notion of mass as "quantity of matter" is mentioned explicitly, apart from the passages quoted above.

Newton's definitions of mass and of inertial force became the subject of numerous comments and often of severe criticism. An interesting remark was made by the Franciscans LeSeur and Jacquier in the 1739 Geneva edition of the *Principia.* With reference to Definition 3 they state that the inertial force is a property of matter, invariant with respect to motion or rest. A transition from rest to motion, they argue, is met with the same opposition on the part of the body as a transition from motion to rest.[20] Phrased in modern terminology, their statement implies that the inertial resistance is invariant under a transformation to a uniformly moving system of reference. With respect to Newton's definition of mass, LeSeur and Jacquier comment that the quantity of matter is an aggregate or the sum of all material particles of which the body is composed. If there were no interstices between the particles, they claim, mass and volume would be the same.[21]

If Newton's definition is not taken as an *ignotum per ignotius,* one has to conclude that for Newton the notion of density was primary and anterior to the concept of mass.[22] This interpreta-

[20] "Vis illa inertiae eadem est in corporibus motis et quiescentibus; tam enim resistunt corpora actioni qua a quiete ad motum concitantur quam actioni qua a motu ad quietem reducuntur. Eadem quippe vis requiritur ad motum datum producendum et ad eundem extinguendum." *Philosophiae naturalis principia mathematica,* ed. Thomas LeSeur and Franciscus Jacquier (Glasgow, ed. 2, 1760), p. 4.

[21] *Ibid.,* p. 1. "In corpore dato materiae quantitatem seu massam, a corporibus magnitudine, aut volumine seu mole distingui oportet. Materiae quantitas est aggregatum, seu summa omnium materiae particularum quibus compositum est corpus . . . Si nulla sint inter solidas corporis partes admixta foramina, massa et volumen non differunt."

[22] Sir William Thomson and Peter Guthrie Tait, in their *Treatise on natural philosophy,* vol. 1, part 1 (Cambridge, 1879), p. 220, remark that "in reality the definition gives us the meaning of density rather than of mass." It is interesting to note that even in the twentieth century some authors still define mass in terms of density and subsequently density in terms of mass. Cf. P. G. Tait and W. J. Steele, *Treatise on the dynamics of a particle* (Macmillan, London, 1900), p. 42; E. J. Andrews and H. N. Howland, *Elements of physics* (Macmillan, London, 1903), p. 11; E. A. Bowser, *An elementary treatise on analytical mechanics* (Van Nostrand, New York, 1904), p. 6.

tion has been advocated by Rosenberger[23] and also by Bloch,[24] who mentions in this connection Boyle's famous experiments on the compressibility of air in which the notion of density played an important role and with which Newton was undoubtedly familiar. Similarly, Crew argues that in Newton's times density, synonymous with specific gravity, length, and time were the three fundamental dimensions in physics, and not mass, length, and time as today. "On such a system, it is both natural and logically permissible to define mass in terms of density."[25] Cohen agrees with Crew's point of view, yet with reservation. According to Cohen, Crew's interpretation would accord with the atomistic view that the fundamental property of a body is the number of corpuscles per unit of volume, or density; but, as Cohen adds, "unfortunately Newton was discreetly silent on this question and such guesses as we may make cannot be confirmed by statements in the *Principia*."[26] Burtt also sees in Boyle's experiments the reason for Newton's use of "density" in defining mass. "Indeed," says Burtt, "having chosen to define it in terms then more familiar rather than present it as an ultimate quality of bodies, he could hardly have done better."[27]

A similar, though not identical, interpretation has been advanced by Hoppe, according to whom Newton assumed that all fundamental particles of which matter is constituted are of the same density and of the same size; the densities of different bodies would thus be proportional to the numbers of particles in equal volumes. Density, consequently a universal constant,

[23] Ferdinand Rosenberger, *Isaac Newton und seine physikalischen Principien* (Leipzig, 1895), part 3, pp. 173, 192.

[24] Leon Bloch, *La philosophie de Newton* (Paris, 1908), p. 140: "Cette idée n'a pas besoin d'être expliquée, Newton la considère comme claire par elle-même et il s'en sert pour définir les autres."

[25] Crew, reference 12, p. 124; ed. 2 (1935), p. 127.

[26] I. Bernard Cohen, *Franklin and Newton* (American Philosophical Society, Philadelphia, 1956), p. 115.

[27] Edwin Arthur Burtt, *The metaphysical foundations of modern science* (Doubleday, New York, 1954), p. 241.

would be more fundamental than mass and prior to it.[28] Although such an interpretation may find some support in Newton's explanation following Definition 2 and may be also implied by LeSeur and Jacquier's commentary,[29] certain passages in Newton's *Opticks,* in particular, are hardly compatible with Hoppe's interpretation, a point raised by Cajori.[30] The following statement, not quoted by Cajori, at the end of the *Opticks,* seems in this context to be of special importance: "God is able to create Particles of Matter of several Sizes and Figures, and in several Proportions to Space, and perhaps of different Densities and Forces."[31] This passage is of importance, not only because it ostensibly contradicts Hoppe's assumption, but also because the peculiar juxtaposition of "Densities and Forces" may indorse a different interpretation which we shall expound later on. "Forces," here, is primarily the *vis inertiae* to which Newton referred only a few lines preceding this quotation: "these Particles have not only a *vis inertiae,* accompanied with such passive Laws of Motion as naturally result from that Force . . ."[32] A more detailed study of the concept of density at the time of Newton will show that Hoppe's contention is untenable.[33]

A completely different approach to interpreting Newton's definition of mass was taken by Enriques.[34] Let G be a group

[28] Edmund Hoppe, *Archiv für Geschichte der Mathematik, der Naturwissenschaften und der Technik* (new series) *11,* 354–361 (1929). Hoppe believes that this concept of density can be traced back to Boyle, Gassendi, Kepler, and Lubin.

[29] "Cum autem motus totius corporis sit aequalis summae motuum singularum Massae partium, seu elementorum, patet manente celeritate, motum totius massae crescere prout crescit numerus elementorum massae aequalium, seu quantitatem motus esse proportionalem massae." Reference 20, p. 3.

[30] See his edition of the *Principia,* reference 15, Appendix, p. 638.

[31] *Opticks* (Dover, New York, 1952), pp. 403–404.

[32] *Ibid.,* p. 401.

[33] For valuable information on the history of the concept of density in the seventeenth century see Helène Metzger's article, "Newton: Sa définition de la quantité de matière et la loi de la conservation de la masse," *Archeion, Archivio di Storia della Scienza 9,* 243–256 (1928).

[34] Federigo Enriques, *Problems of science,* trans. K. Royce (Open Court, Chicago; 1914), pp. 269–271.

of mechanical operations comprising movements, divisions, recompositions, compressions, and dilatations. Newton's objective, according to Enriques, was to find an additive invariant with respect to G. Let us assume that parts of any two given bodies can be made "equal," in the sense that one can be substituted for the other, by transformations of G. A standard body A is chosen, for every part of which the mass is defined as proportional to volume. For any given homogeneous body B, then, the density is the inverse ratio of its volume to that of a transformed element equal to a part of A. In this sense, mass is eventually defined as the product of volume and density. Enriques' vindication of Newton's definition, although logically unimpeachable, sounds too artificial and too much *ad hoc* to be acceptable as historically true.

In contrast to these attempts to find some logical justification, Volkmann bluntly rejects Newton's definition of mass, censuring it as circular since density, in its turn, is definable only as mass per unit volume.[35] Volkmann remarks: "Newton said: 'I do not define time, space, place, and motion, as being well known to all.' Newton equally well could have declared: 'I do not define mass.' "[36]

Volkmann, in this criticism of Newton, followed Ernst Mach, who in *The science of mechanics*[37] describes Newton's definition as "unfortunate" (*unglücklich*). Mach's strong opposition to Newton's definition is also directed, apart from the circularity involved, against the use of terms such as "quantity of matter," which for Mach have no physical significance at all. In fact, it must be admitted that prior to Mach's critique, the terms "quantity of matter," "mass," and so forth were used in a most uncritical way. Thus, for instance, Henry, Lord Brougham, a

[35] Paul Volkmann, *Erkenntnistheoretische Grundzüge der Naturwissenschaften und ihre Beziehungen zum Geistesleben der Gegenwart* (Teubner, Leipzig and Berlin, ed. 2, 1910), p. 359.

[36] *Ibid.*, p. 126.

[37] Ernst Mach, *Die Mechanik in ihrer Entwicklung* (Leipzig, ed. 1, 1883), trans. T. J. McCormack, *The science of mechanics* (Open Court, LaSalle, Ill., 1942), chap. 2, sec. 3.

Fellow of the Royal Society, wrote in his Commentary to the *Principia*: "The *Principia* begins with a definition of terms . . . The definitions, eight in number, comprise that of quantity of matter, which is in the proportion of its bulk and density, the density being the proportion of its mass to its bulk." [38]

The problem of Newton's definition of mass may, of course, be explained by stating that Newton simply repeated Kepler's idea, as expressed in the *Prodromus dissertationum cosmographicarum continens mysterium cosmographicum*,[39] where Kepler refers to the volume of the body and to the density of its matter as the constituent factors of inertia. A more critical analysis of Newton's writings, however, seems to suggest a different explanation. For two fundamental, and at first independent, notions seem to have dominated Newton's conception of matter: *quantitas materiae* and *vis inertiae*. In his Rules of Reasoning, at the beginning of the "System of the world," Book III of the *Principia,* Newton characterizes as universal qualities of all bodies those qualities that "admit neither intensification nor remission of degrees, and which are found to belong to all bodies within the reach of our experiments." [40] Extension, impenetrability, and mobility exist in the whole body because they exist in the parts; they are properties whose range of applicability, so to speak, extends both to fundamental particles and to macroscopic bodies composed of the latter. The same holds, according to Newton, for inertia, whereas gravity, as diminishing with distance from the central body, is not reckoned among the universal properties of matter. "Not that I affirm gravity to be essential to bodies: by their *vis insita* I mean nothing but their inertia. This is immutable." [41]

Extension, impenetrability, and mobility had a similar func-

[38] Henry, Lord Brougham, *Analytical view of Sir Isaac Newton's Principia* (London, 1855), p. 13.

[39] Kepler, *Opera omnia,* ed. C. Frisch (Frankfort and Erlangen), vol. 1 (1858), p. 161.

[40] Reference 15, p. 398.

[41] *Ibid.,* p. 400.

tion in Descartes's theory of matter; hardness was essential for atomistic philosophy. But how did an innate force, the *vis insita* or *vis inertiae,* become a universal property of matter? Although this "force of inactivity," as Newton explicitly states, is exerted only when another force endeavors to change the condition of the body, the force as such exists in the body as if hibernating through all time. The requirement of its existence is historically conditioned. Pre-Galilean physics, as is well known, could not formulate a law of inertia and its discovery had to wait until the times of Beeckmann, Gassendi, and Descartes, partly because of its conceptual incompatibility with the scholastic dictum: *cessante causa cessat effectus.* With the exception of a few authorities, such as Alhazen or Olivi, medieval philosophy viewed causal processes as instantaneous and regarded the evanescence of the cause as leading to an immediate evanescence of the effect. Motion, consequently, presupposes a *vis motrix* throughout the duration of its existence.

In Newton's mechanics, for the case of uniform motion, it is the *vis inertiae* that plays, though tacitly, the role of a *vis motrix.* In fact, it is not only resistance to change of motion, it is also "the power of persevering in motion,"[42] And yet, it is not force in the conventional sense of the word. In scholastic philosophy a force without resistance can never give rise to a (finite) motion. In Newtonian mechanics a force without a counterforce, according to the action-reaction principle, is inconceivable. It thus assumes the character of a real force only in changes of motion. Different bodies, as Newton realizes, evoke different resistances. Now, classical, or what is usually called Newtonian, mechanics takes this resistance to change of motion, different in different bodies but a constant for the same body, as the ultimate and absolute individual characteristic of the given body; "inertial mass," as this absolute parameter is called, thus becomes for a given body an irreducible magnitude, on which other parameters depend, but which itself is independent of anything else. Not so

[42] *Ibid.*, p. 399.

for Newton! Newton postulated a proportionality between *vis inertiae* and another fundamental characteristic of a given body, its *quantitas materiae.*[43] Thus for Newton, in contrast to "Newtonian mechanics," "inertial mass" is a reducible property of physical bodies, depending on their "quantity of matter." For Newton the concept of *quantitas materiae* is still a concept of physical significance. Now, since bodies of equal volumes possess, in general, different inertial forces, their "quantities of matter" have also to be different. An "intensive," in the sense of volume-independent, factor has thus to be identified as responsible for the difference of *quantitas materiae* in bodies of equal volumes, in general. And this factor, because of the property of inertia of being universal, has also to be a characteristic of the smallest particles conceivable. This factor, in a purely intensive, qualitative sense, and irreducible just like the notion of affinity in classical chemistry, is identified by Newton as "density."

It is, of course, always rather difficult to state exactly whether two concepts are involved or only one, once a general proportionality between the concepts has been established. In the present case, however, it is fairly obvious that "quantity of matter" is still a notion for itself. A careful examination of the text of Book III of the *Principia,* for instance, shows clearly that it is *quantitas materiae* in the original sense of the word which determines the magnitude of gravitational attraction.[44] It has also been suggested that Newton, in postulating the proportionality between *quantitas materiae* and *vis inertiae,* was influenced by Baliani's *De motu gravium,*[45] but little evidence can be adduced.

That gravity or weight is proportional to *quantitas materiae* for a given locality was shown by Newton in a series of experiments which he describes in Book III, Proposition 6, Theorem 6,

[43] See Newton's explanation after Definition 3, quoted on p. 65.

[44] "All bodies . . . gravitate . . . in proportion to the quantity of matter which they severally contain." Reference 15, p. 399.

[45] See p. 62. This conjecture was made by J. M. Child in his essay "Newton and the art of discovery," in *Isaac Newton: 1642–1727,* ed. W. J. Greenstreet (Bell, London, 1927), p. 127.

of the *Principia*.[46] Of two equal wooden boxes he filled one with wood and suspended an equal weight of gold in the center of oscillation of the other. Using these boxes as pendulums with equal lengths of suspension he noticed that their periods were equal. "I observed them to play together forwards and backwards, for a long time, with equal vibration," he reports. "And therefore," he adds, "the quantity of matter in the gold was to the quantity of matter in the wood . . . as the weight of the one to the other." Newton's reasoning, expressed in modern pre-Einsteinian terminology, was as follows: the period T for a pendulum of length L in a region characterized by the constant of free fall g is given by the formula $T = 2\pi(L/g)^{1/2}$; on the other hand, the weight W and the quantity of matter m are related by $W = mg$. Thus, $T = 2\pi(mL/W)^{1/2}$ or $W/m = 4\pi^2 L/T^2$. Since the experiment shows that the measurable ratio L/T^2 is independent of whether wood, gold, lead, glass, sand, or wheat is used, the ratio W/m, that is the ratio between weight and quantity of matter, is a constant. Newton concludes the description of his experiments with the remark that "a difference of matter less than the thousandth part of the whole" would have easily been detected.

These experiments, which were repeated with more refined instrumentation in 1832 by Friedrich Wilhelm Bessel,[47] justify the use of the balance for the comparison or determination of mass.

That in spite of this proportionality weight and mass are totally different concepts was experimentally shown in 1671 by Jean Richer. Charged by the French Academy to proceed to Cayenne in French Guiana in order to carry out astronomical observations for the determination of the solar parallax and related measurements, he found that his pendulum clock lost daily about

[46] Reference 15, p. 411.

[47] Friedrich Wilhem Bessel, "Studies on the length of the seconds pendulum," *Abhandlungen der Preussischen Akademie* (1826); "Experiments on the force with which the earth attracts different kinds of bodies," *ibid.* (1830). Cf. also *Poggendorffs Annalen der Physik und Chemie 25,* 1–14 (1832) and *26,* 401–411 (1833). Among the substances tested by Bessel were meteoric iron and meteoric stones.

2½ minutes with respect to mean solar time.[48] Newton refers to these observations in his *Principia* under the heading: "to find and compare together the weights of bodies in the different regions of our earth."[49] "Weight," which till this time was generally regarded as a fundamental and often constitutive attribute of matter, revealed itself now as reducible to mass and acceleration. Quantity of matter, or mass, and weight were from now on two distinctly different concepts. The technical designation of the first as "mass" and its sharp contradistinction from "weight," although implicitly contained in Newton's *Principia,* were for the first time clearly emphasized by John Bernoulli in his *Meditatio de natura centri oscillationis,*[50] in which he stated that "mass" times the acceleration of free fall is the weight of the body.

[48] "Observations . . . par Monsieur Richer de l'Académie Royale des Sciences, *Mémoires,*" vol. 7, part 1 (Paris, 1729), pp. 233–326.

[49] Reference 15, p. 428.

[50] "Ex ratione materiae quantitatis, quam vocabo massam vel molem, et ex ratione gravitatum acceleratricium; componendo namque duas posteriores, nascitur ratio ponderum." *Opera omnia* (Lausanne and Geneva, 1742), vol. 2, p. 169.

CHAPTER 7

PHILOSOPHICAL MODIFICATIONS OF THE NEWTONIAN CONCEPT

Newton's systematization of the concept of mass and the clarification of its relation to that of weight were of decisive importance for the development not only of mechanics and physics, but of science in general. In mechanics, mass became the fundamental characteristic of bodies and eventually identical with "body." "The definition of physical bodies as masses was the signal achievement needed in modern mechanics after the nature of space had been discovered by Galileo and Descartes and that of time formulated by Barrow." [1]

In chemistry the systematization of mass contributed to the downfall of alchemy and gave support to the rise of modern quantitative chemistry. It will be recalled that "weight" for the alchemist was an accidental property of matter: an increase of weight was not necessarily associated with an influx of additional matter. In other words, the *quantitas materiae* did not necessarily change if, for instance, lead became gold and increased in weight in the course of this transmutation. The alchemists were convinced that a small quantity of *magisterium* could transmute a few pounds of lead into hundreds of pounds of gold.[2] Only oc-

[1] Edwin Arthur Burtt, *The metaphysical foundations of modern science* (Doubleday, New York, 1954), p. 240.

[2] Johannes Wislicenus, *Die Chemie und das Problem der Materie* (Leipzig, 1893), p. 24.

casionally had the use of the balance been recommended as an instrument of research. Nicholas of Cusa, in his *De staticis experimentis*,[3] sees in weighing the most reliable and most accurate method of research, for God set the world in order "in mensura et numero et pondere."[4] But Cusanus was an exception. The use of the balance as a systematic method in chemical or physical research obtained its scientific foundation and logical justification only through Newton's work, and in particular, through the established proportionality between mass and weight.

That such conclusions were not unconfuted even at a considerable time after the publication of the *Principia* can be seen if one studies, for instance, Leibniz's theory of matter. Still in 1716, in his fifth letter to Clarke, Leibniz says: "And as for quicksilver; 'tis true, it contains about fourteen times more of heavy matter, than an equal bulk of water does; but it does not follow, that it contains fourteen times more matter absolutely."[5]

Leibniz's theory of matter, and his concept of mass, in particular, are most complicated. Couturat, in an interesting article, "Le système de Leibniz d'après M. Cassirer,"[6] declared that Leibniz invented the concept of mass.[7] Mach, on the other hand, flatly denies that Leibniz ever was in possession of a concept of mass.[8] How can two such acute scholars as Couturat and Mach so obviously contradict each other?

Leibniz's original concept of mass was fundamentally different

[3] Nicholas of Cusa, *Opera omnia*, ed. C. Baur (Leipzig, 1937).

[4] *Ibid., Liber sapientiae,* chap. 11, sec. 21.

[5] *The Leibniz-Clarke correspondence,* ed. H. G. Alexander (Philosophical Library, New York, 1956), p. 66.

[6] Louis Couturat, *Revue de métaphysique et de morale 11,* 83–99 (1903).

[7] "L'invention du concept de masse ne constituait pas seulement un progrès capital de la mécanique: elle permettait à Leibniz de dissocier complètement l'idée de matière de l'idée d'étendue, puisque le coefficient appelé masse est une quantité numérique, et non une grandeur spatiale."

[8] "Einen eigentlichen Massenbegriff hat Leibniz so wenig als Descartes; er spricht vom Körper (corpus), von der Last (moles), von ungleich grossen Körpern desselben spezifischen Gewichtes usw. Nur in der zweiten Abhandlung (1695) kommt einmal der Ausdruck 'massa' vor, welcher wahrscheinlich Newton entlehnt ist." Ernst Mach, *Die Mechanik* (ed. 8, Brockhaus, Leipzig, 1921), p. 289.

from the Newtonian notion. In a letter, of 1669, to Jacob Thomasius, his favorite teacher at the University of Leipzig, Leibniz wrote: "Primary matter is mass itself in which there is nothing but extension and antitypy or impenetrability."[9] Leibniz's theory of matter can be understood only if viewed against his doctrine of monads. Primary matter (*materia prima*), according to Leibniz, is not body but intrinsic to the "being a monad." Secondary matter (*materia secunda*), on the other hand, does not refer to the monads themselves but belongs to a group, or cluster, of monads; it is founded on the ontological relation between the subordinate monads and the dominant monads; the subordinate monads, reflecting by their nature with lower degrees of clearness, are also the more passive ones. These, in turn, have also secondary matter through the subordination of other monads to them, and so on ad infinitum. Completely inert matter thus becomes a limiting concept.

Leibniz now applies the same terms of primary and secondary matter also to the world of perception, the world as the object of physical research and not of metaphysical scrutiny. Primary matter, or mass as Leibniz calls it in his letter to Thomasius, is here an abstraction; it is body conceived as merely occupying space and preventing other bodies from occupying the same space. Extension and antitypy, a favorite term with Leibniz for impenetrability,[10] are thus the attributes of *materia prima*. On other occasions Leibniz reserves the term mass for secondary matter, a

[9] "Materia prima est ipsa Massa in qua nihil aliud quam extensio et antitypia seu impenetrabilitas." Letter to Jacob Thomasius, April 1669, in C. I. Gerhardt, ed., *Die philosophischen Schriften von G. W. Leibniz* (Berlin, 1880), vol. 4, p. 165. Mach's contention, as far as the date (1695) is concerned, is, as we see, erroneous.

[10] "Materia in se sumta seu nuda constituitur per Antitypiam et Extensionem." *Commentatio de anima brutorum, Opera philosophica quae extant, latine, gallica, germanica omnia,* ed. J. E. Erdmann (Berlin, 1840), part 1, p. 463. In a letter to Wagner (Epistola ad Wagnerum de vi activa corporis, 1710) Leibniz wrote: "Principium activum non tribui a me materiae nudae sive primae, quae mere passiva est, et in sola antitypia et extensione consistit." *Ibid.,* p. 466. The term "antitypia," as used originally by Plutarch and Sextus Empiricus, meant at that time only "hardness."

concept less abstract than primary matter because of the added notion of activity. Thus in a letter to John Bernoulli Leibniz characterizes primary matter by *moles* and secondary matter by *massa:* "Matter in itself, or *moles,* which may be called primary matter, is not a substance, nor even an aggregate of substances, but something incomplete. Secondary matter, or *massa,* is not one substance, but a plurality of substances."[11] Mass is an aggregation of several substances, like a pondful of fish or a flock of sheep. It is an attribute of a collection and as such a *unum* only *per accidens,* somewhat like the concept of space, which according to Leibniz is formed by hypostatizing a system of relations and endowing it with ontological existence.[12]

Mass, although initially merely a *phenomenon bene fundatum,* plays an important role in the world of perception. For the behavior of physical bodies cannot be understood unless, apart from extension and impenetrability, something else is assumed to belong to bodies, some quality in virtue of which "a big body is more difficult to be set in motion than a small body."[13] Leibniz, fully aware of the fundamental conceptual deficiency of Cartesian physics, thus attributes inertia to mass qua *materia secunda.* Geometry alone cannot account for the spatio-temporal behavior

[11] "Materia ipsa per se, seu moles, quam materiam primam vocare possis, non est substantia; imo nec aggregatum substantiarum, sed aliquid incompletum. Materia secunda, seu massa, non est substantia, sed substantiae; ita non grex, sed animal; non piscina, sed piscis, substantia una est." *Leibnizens mathematische Schriften,* ed. G. I. Gerhardt (Halle, 1856), vol. 3, part 2, p. 537. The same distinction in terms is found in his "De ipsa natura, sive de vi insita actionibusque creaturarum," *Opera omnia,* vol. 2, part 2, p. 54.

[12] Cf. Max Jammer, *Concepts of space* (Harvard University Press, Cambridge, 1954), pp. 114–117. Leibniz's conception of mass as a class concept reappears in Roger Joseph Boscovich's kinematic theory, where mass is a function of the structure of the system. See L. L. Whyte, "Boscovich and particle theory," *Nature 179,* 284–285 (1957).

[13] "Nous remarquons dans la matière une qualité que quelques-uns ont appellée l'inertie naturelle, par laquelle le corps résiste en quelque façon au mouvement; en sorte qu'il faut employer quelque force pour l'y mettre, (faisant même abstraction de la pesanteur) et qu'un grand corps est plus difficilement ébranlé qu'un petit corps." *Journal des sçavans* (18 juin 1691), p. 259. Cf. *Opera,* reference 10, p. 142.

of bodies interacting with each other.[14] Inertial mass, in conformity with physical experience, thus becomes for Leibniz a conceptual necessity which has to be posited so that the equality of cause and effect is empirically assured.

It is difficult to assess to what extent Leibniz's concept of mass, as outlined so far, has been directly influenced by Newton's writings. To Kepler, whom he mentions repeatedly in this context, he certainly owes much of his physical reasoning. Leibniz's concept of mass, as presented in his *Dynamica de potentia et legibus naturae corporea* (1689), which he began to write during his visit to Rome, resembles more and more that of Newton.[15] In fact, Leibniz's definition of mass, at this stage, in spite of his declared opposition to Newton's theory of gravitation, is wholly Newtonian: "The masses of mobile bodies are to each other as their volumes and densities, or as the extension and intension of matter."[16]

Leibniz always insisted that his physics should accord and harmonize with his theological doctrines, yet without assuming any mutual interpenetration. No spiritual principles should be applied for the explanation of physical phenomena nor should physical laws be used for the interpretation of religious events. Showing that a religious occurrence is possible, that it is logically compatible with the laws of physics, does not mean to explain it by physics. If Leibniz, consequently, employs the notion of inertial mass, the inherent quality of matter additional to extension and impenetrability, in order to show the physical possibility of

[14] "S'il n'y avoit dans les corps que l'étendue ou la situation, c'est-à-dire ce que les Géomètres y connoissent, joint à la seule notion du changement, cette étendue seroit entièrement indifférente à l'égard de ce changement; et les résultats du concours des corps s'expliqueroient par la seule composition Géométrique des mouvements . . . ce qui est entièrement irréconciliable avec les expériences." *Mathematische Schriften,* reference 11, vol. 6, p. 240.

[15] It was in Rome that through the *Acta eruditorum* Leibniz got his first information on the *Principia.*

[16] "Moles mobilium, vel ipsa mobilia, sunt in ratione composita voluminum et densitatem, seu extensionum et intensionum materiae." *Dynamica,* part 1, chap. 2, proposition 3; cf. *Mathematische Schriften,* vol. 6, p. 298.

the Eucharistic mystery, he does not violate his principles. "In revealed theology I attempt to demonstrate, not the factual truth, but the possibility of the mysteries . . . the possibility of the Trinity, the Incarnation, and the Eucharist." [17]

Extension, Leibniz claims, presupposes something extended, something expanded and continued. In milk, it is whiteness; in the diamond, it is hardness; in bodies in general there is thus something anterior to extension: it is action and motion. Being able to exert resistance, even if only against change of motion, mass itself must be a source of power, a dynamical entity.[18] Thus mass, in the final stage of Leibniz's system, becomes the carrier or diffusor of activity and energy.

When reviewing, in retrospect, the various phases of Leibniz's concept of mass in their formulations corresponding to the different aspects of his philosophical system and its evolution from extensionless atomism to energetic dynamism, it is not difficult to understand that for the philosopher it is a matter of great originality and for the scientist a matter of methodological deficiency. The contradiction between Couturat and Mach can thus be rendered intelligible.

Leibniz's dynamical interpretation of mass led to at least one important result for physical science: it accentuated the problematics concerning the logical status and the physical significance

[17] "In Theologia Revelata übernehme ich zu demonstriren, nicht zwar veritatem (denn die flaust a revelatione) sondern possibilitatem mysteriorum, contra insultos infidelium et Atheorum, dadurch sie von allen contradictionibus vindicirt werden, nehmlich possibilitatem Trinitatis, incarnationis, Eucharistiae . . . Ich aber bin endlich durch tieffe untersuchungen dahin kommen, dass ich possibilitatem mysteriorum Eucharistiae, wie sie in Concilio Tridentino ercläret werden, salva philosophiae emendata, welches vielen unglaublich vorkommen wird, zu demonstriren mir getraue. Ich will weissen vi principiorum philosophiae emendatae necesse est, ut datur in omni corpore principium intimum incorporeum substantiale a mole distinctum, et hoc illud esse, quod veteres, quod Scholastici substantiam dixerint, etsi nequiverint se distincte explicare multo minus sententiam suam demonstrare." *Philosophische Schriften,* reference 9, vol. 1, p. 61. Cf. also Leibniz's letter to Arnauld, *ibid.,* p. 75.

[18] Cf. Max Jammer, *Concepts of force* (Harvard University Press, Cambridge, 1957), chap. 9.

of the somewhat paradoxical notion of *vis inertiae* and thus led eventually to the elimination of this concept by Immanuel Kant.

For Newton mass was the carrier of *vis inertiae,* and *quantitas materiae* was proportional to it. This concept of *vis inertiae* was in the seventeenth and eighteenth centuries not a mathematical fiction or an artificial device, as we use it today in connection with d'Alembert's principle[19] or in connection with the inertial behavior of ponderable matter under transformations of coordinate systems. It was a physical existent of ontological reality comparable with any other known physical force and it played an important role at that time in treatises on mechanics.

Kant's examination of the ontological and methodological status of *vis inertiae* was a major contribution to foundational research in physics. It is surprising to see that in spite of the exuberant literature on all possible aspects of Kant's philosophy this important contribution of Kant's seems never to have been called attention to. The problem of mass and its relation to inertia is the main subject in four of Kant's publications on natural philosophy, three of which belong to his precritical period.

In *Thoughts on the true estimation of living forces,*[20] the first of these writings, published in 1747, Kant, claiming to follow Aristotle and Leibniz, ascribes to every physical body inherent forces prior to spatial extension. "There would be no space and no extension, if substances had not force whereby they act outside themselves. For without a force of this kind there is no connection, without this connection no order, and without this order no space." [21] In the *Physical monadology* which still betrays the disciple of Leibniz and Wolff, Kant reduces the impenetrabil-

[19] In the modern formulation of d'Alembert's principle $-d(m\dot{r})/dt$ is usually called "the force of inertia." The combined effect of these inertial forces and the applied forces is known as the effective force whose total work is zero for a virtual displacement compatible with the constraints of the system.

[20] Immanuel Kant, *Gedanken von der wahren Schätzung der lebendigen Kräfte,* in part translated into English by John Handyside in *Kant's inaugural dissertation and early writings on space* (Open Court, Chicago and London, 1929).

[21] *Kant's inaugural dissertation, ibid.,* p. 10.

ity of matter (proposition 8) and its inertia (proposition 11) to forces inherent in matter. Every element of a physical body possesses inertial force to an extent that varies from element to element.[22] In addition 1 to proposition 11, Kant still describes *vis inertiae* as a particular kind of *vis motrix,* a motive force, and in addition 2 he identifies "mass" as the magnitude of this force.[23]

Two years later, however, in the *Neuer Lehrbegriff der Bewegung und Ruhe* (1758), Kant raises the first objection against the legitimacy of the concept of *vis inertiae.* Newton, in his explanation to Definition 3 of the *Principia,* it will be recalled, takes *vis inertiae,* in the case of impact between two bodies, as an opposition of the "inert" body against the impressed force. Kant, who followed Newton so far, now raises the interesting question: how can the equilibrium of the intrinsic forces of a body be suddenly disturbed as soon as the moving body touches the body at rest, and just in such a way that a force emerges that acts in the direction opposing that of the approaching body. In attempting to reduce the phenomenon under discussion to the principle of action and reaction, whereby the body "at rest" must be regarded as moving relatively to the approaching body,[24] Kant concluded that *vis inertiae,* although a notion convenient for the formulation and deduction of the laws of motion, is fundamentally a

[22] "Vi inertiae cuiuslibet elementi alia vel maior vel minor dari poterit in diversae specei elementis," *Metaphysicae cum Geometria junctae usus in Philosophia naturali specimen primum continet Monadologiam physicam* (Königsberg, 1756); Kant's *Werke* (Akademieausgabe; Berlin, 1910), vol. 1, p. 485.

[23] "Massa corporum non est nisi ipsorum vis inertiae quantitas, qua vel motui resistunt vel data celeritate mota certo movendi impetu pollent." *Ibid.*

[24] "Nun ich aber bewiesen habe, dass, was man fälschlich vor eine Ruhe in Ansehung des stossenden Körpers gehalten hat, in der That beziehungsweise auf ihn eine Bewegung sei, so leuchtet von selber ein, dass diese Trägheitskraft ohne Noth erdacht sei, und bei jedem Stosse eine Bewegung eines Körpers gegen einen andern, mit gleichem Grade ihm entgegen bewegten angetroffen werde, welches die Gleichheit der Wirkung und Gegenwirkung, ohne eine besondere Art der Naturkraft erdenken au dürfen, ganz leicht und begreiflich erkläret. Gleichwohl dienet diese angenommene Kraft ungemein geschickt dazu, alle Bewegungsgesetze sehr richtig und leicht daraus herzuleiten." *Neuer Lehrbegriff der Bewegung und Ruhe,* in *Immanuel Kants kleinere Schriften zur Naturphilosophie,* ed. O. Bük (Meiner, Leipzig, 1907), p. 402.

superfluous and unnecessary concept. Finally, in the *metaphysical foundations of natural science,*[25] published five years after the *Critique of pure reason,* Kant rejects Newton's concept of *vis inertiae* altogether, and this in spite of the fact that the philosophical foundation of Newton's mechanics was the very objective of this work. As only "motion can oppose motion, but not rest," [26] it is not the inertia of matter, its incapability to move itself, that produces resistance to the moving force. A force which by itself does not cause motion, but only resistance, is "a word without any meaning." [27] The concept of a *vis inertiae* has to be abandoned in natural science, Kant concludes, not merely because of its paradoxical appellation, but because of the misconceptions which the term implies, and this "notwithstanding the famous name of its originator." Instead of the "force of inertia" Kant postulates the "law of inertia" as corresponding to the category of causality: every change of the state of motion has an external cause.[28] What now happens to the concept of mass which Kant originally identified as the magnitude of this force? "Quantity of matter" is now, according to Kant, the amount (*Menge*) of the mobile (*Beweglichen*) in a definite volume and "mass" is "quantity of matter" regarded as active together at the same time.[29] Such a definition would be meaningless unless it were stipulated how to measure this quantity. In proposition 1 Kant declares explicitly: "The quantity of matter, in comparison with any other, can be estimated only by means of the quantity of mo-

[25] Immanuel Kant, *Die metaphysischen Anfangsgründe der Naturwissenschaften* (Riga, 1786), translated into English by Ernest Belfort Bax in *Kant's Prolegomena and Metaphysical foundations of natural science* (London, 1909).

[26] "Einer Bewegung kann nichts widerstehen, als entgegengesetzte Bewegung eines anderen, keineswegs aber dessen Ruhe," reference 24, p. 301.

[27] "Ein Wort ohne alle Bedeutung," *ibid.,* p. 302.

[28] "Alle Veränderung der Materie hat eine äussere Ursache. Ein jeder Körper verharrt in seinem Zustande der Ruhe oder Bewegung . . . wenn er nicht durch eine äussere Ursache genötigt wird, diesen Zustand zu verlassen." *Ibid.,* p. 291.

[29] "Die Quantität der Materie ist die Menge des Beweglichen in einem bestimmten Raum. Dieselbe, sofern alle ihre Teile in ihrer Bewegung als zugleich wirkend (bewegend) betrachtet werden, heisst die Masse." *Ibid.,* p. 283.

tion with given velocity."[30] Although Kant's presentation of these ideas is sometimes rather obscure and the modern reader may look in vain for additional information of operational significance, it is obvious that Kant was fully aware of the problematic character of Newton's concept of mass. Kant's elimination of the metaphysical *vis insita* or *vis inertiae* prepared the way for a more positivistic approach to the concept of mass.

[30] "Die Quantität der Materie kann in Vergleichung mit jeder anderen nur durch die Quantität der Bewegung bei gegebener Geschwindigkeit geschätzt werden." *Ibid.,* p. 284.

CHAPTER 8

THE MODERN CONCEPT OF MASS

The natural philosophy of the eighteenth and nineteenth centuries was dominated by what has been called the substantial concept of matter: material objects were regarded as containing a substantial substratum underlying all physical reality. This substratum, furthermore, was described as absolute since it functions as the carrier of the changing sensory qualities without being itself affected by these qualities, just as Newtonian space was absolute as acting as an inertial system on all material objects without being acted upon by these objects. The preservation, in time, of the identity of physical objects, in spite of the constantly shifting sensory qualities, was relegated to the substantiality of matter.

With the gradually increasing recognition of inertial mass these metaphysical considerations appeared to obtain a scientific foundation in the principle of conservation of mass. The idea that the quantity of matter is preserved in the course of the history of a material system was, of course, an implicit assumption already in the *Principia*. Its methodological importance was emphasized by Kant, who placed the principle on an equal footing with the laws of motion,[1] with the remark that "general metaphysics" posits this principle as a postulate. In fact, it was

[1] Immanuel Kant, *Metaphysical foundations of natural science,* proposition 2, in *Immanuel Kants Kleinere Schriften zur Naturphilosophie,* ed. O. Bük (Meiner, Leipzig, 1907).

regarded as being essentially a restatement of the ancient idea of the indestructibility and uncreatedness of matter as expressed by the early atomists, Empedocles, Aristotle, Lucretius, and others.

New support to the principle was given in the eighteenth century through the chemical investigations of Antoine Lavoisier. In his *Traité élémentaire de chimie,* published in 1789,[2] Lavoisier showed that the principle of conservation of mass applies also to chemical reactions. "Nothing is created in the course of artificial or natural reactions, and it can be taken as an axiom that in every reaction the initial quantity of matter is equal to the final quantity of matter."[3] Lavoisier exemplified the principle by the process of vinous fermentation in which the weight or mass of the materials, water, sugar, yeast, present before the fermentation is equal to the weight of the components after the fermentation (residual liquor, carbonic acid gas). Subsequent to the publication of this result, numerous other chemical reactions were quantitatively investigated with the use of the balance, and the principle was readily accepted by the early propagators of modern chemistry. Since the variation of weight with change of position had meanwhile become common knowledge, it was clear that it was actually mass and not weight that was conserved in physical or chemical operations. Since, furthermore, mass was a universal, and the only universal, characteristic of physical bodies that was conserved, the concept of mass was for all practical purposes identified with that of substance. It thus became the dominating notion of the substantial concept of matter. Still in 1896 Charles de Freycinet, in his essays on the philosophy of science, declared: "If I should have to define matter I would say: matter is all that has mass, or all that requires force in order to be set in motion."[4]

Yet, in spite of its dominating role, no formal definition of the

[2] The first English translation of the *Traité élémentaire de chimie,* by Robert Kerr, appeared in 1790 in Edinburgh under the title *Elements of Chemistry in a new systematic order, containing all the modern discoveries.*

[3] *Ibid.,* chap. 13, p. 101.

[4] "Si j'avais à définir la matière, je dirais: la matière est tout ce qui a de la masse, ou tout ce qui exige de la force pour acquérir du mouvement." Charles de Freycinet, *Essais sur la philosophie des sciences* (Paris, 1896), p. 168.

concept could be given. Usually one took it synonymous with "quantity of matter" without specifying how to measure the quantity and without supplying any other operational interpretation. Thus, the *Dictionnaire raisonné de physique* by Brisson defines "mass" as the quantity of matter which a body contains.[5] The *Dictionnaire de physique,* published by Sigaud de la Fond, gives an almost identical definition.[6] Textbooks or treatises on mechanics in the eighteenth century propose no better definition of mass.

The only exception is Leonard Euler's *Mechanica,* written with the specific purpose of constructing, by axioms, definitions, and logical deductions, a rational science of mechanics, by which Euler tried to demonstrate the apodictic character of Newtonian mechanics. For the history of the concept of mass Euler's *Mechanica* is of outstanding importance, for it constitutes the logical transition from the original Newtonian conception of mass, based on the concept of *vis inertiae,* to the more modern abstract conception as a numerical coefficient, characteristic of the individual physical body and determined by the ratio of force to acceleration.

Proposition 7 of the first book of the *Mechanica* states the law of inertia for bodies at rest which, according to Euler, can be demonstrated by the principle of sufficient reason. "For there is no reason whatever that a body should move in this rather than in that direction."[7]

[5] "Masse. On appelle ainsi, en physique, la quantité de matière propre que contient un corps." M. Brisson, *Dictionnaire raisonné de physique* (Paris, 1781), vol. 2, p. 108.

[6] "On entend en physique par la masse d'un corps la quantité de matière qu'il contient." Sigaud de la Fond, *Dictionnaire de physique* (Paris, 1781), vol. 3, p. 121.

[7] "Theorema: Corpus absolute quiescens perpetuus in quiete perseverare debet, nisi a causa externa ad motum sollicitetur. Demonstratio: Concipiamus corpus hoc existere in spatio infinito atque vacuo, perspicuum est nullam esse rationem, quare potius in hanc vel illam plagem moveatur. Consequenter ob defectum sufficientis rationis, cur moveatur, perpetuo quiescere debebit." Leonhard Euler, *Mechanica sive motus scientia analytice exposita* (St. Petersburg, 1736), ed. Paul Stäckel (Leipzig and Berlin, 1912), vol. 1, p. 27. The principle of sufficient reason was repeatedly applied for a "proof" of the law of

The second proposition of interest for us is proposition 17: "The force of inertia in any body is proportional to the quantity of matter which the body contains." This, of course, is a restatement of Newton's explanation to Definition 3 in the *Principia.* So far Euler just followed Newton. But in the proof of proposition 17 a new idea turns up. Euler describes the *vis inertiae,* the force in virtue of which the body perseveres in its state of rest or uniform motion, as determined by the force (*vi vel potentia*) that is required to dislodge (*deturbare*) the body from its state of rest or motion. Different bodies require different forces in proportion to their quantities of matter.[8] Thus the "quantities of matter" or "masses" are determined by motive forces, an idea which in the subsequent chapters of the *Mechanica* as well as in his *Theoria motus corporum solidorum seu rigidorum* (1760) gained increasing preponderance. Thus in Corollarium 2 to his definition of

---

inertia, both for bodies at rest and for bodies in uniform motion. See our remarks on the Brethren of Purity (p. 32), and also Aristotle, *Physics,* book 4, chap. 8, 215 a 19 (argument against motion in a vacuum); J. le Rond d'Alembert, *Traité de dynamique,* part 1, chap. 1, law 2; I. Kant, *Metaphysical foundations,* part 3, proposition 3; R. H. Lotze, *System der Philosophie* (*Metaphysik*) (Leipzig, 1874), p. 311; A. Schopenhauer, *Welt als Wille und Vorstellung* (Leipzig, 1877), vol. 1, p. 79; J. C. Maxwell, *Matter and motion,* chap. 3 (Dover, New York), p. 28; K. Kroman, *Unsere Naturerkenntnis* (Copenhagen, 1883); E. Dühring, *Logik und Wissenschaftstheorie* (Leipzig, 1878), p. 280. For criticisms of such proofs see: F. Poske, "Der empirische Ursprung und die Allgemeingültigkeit des Beharrungsgesetzes," *Vierteljahrschrift für wissenschaftliche Philosophie 8,* 385–404 (1884); P. Johannesson, *Das Beharrungsgesetz* (Berlin, 1896); P. Frank, *Philosophy of science* (Prentice-Hall, Englewood Cliffs, New Jersey, 1957), pp. 166–168; W. Wundt, *Die physikalischen Axiome und ihre Beziehung zum Causalprincip* (Erlangen, 1866), p. 40; H. Streinitz, *Die physikalischen Grundlagen der Mechanik* (Leipzig, 1883), p. 53; and K. Lasswitz, *Atomistik und Kriticismus* (Braunschweig, 1878), p. 78.

[8] "Theorema: Vis inertiae cuiuscunque corporis proportionalis est quantitati materiae, ex qua constat. Demonstratio: Vis inertiae est vis in quovis corpore insita in statu suo quietis vel motus aequabilis in directum permanendi. Ea igitur aestimanda est ex vi vel potentia, qua opus est ad corpus ex statu suo deturbandum. Diversa vero corpora aequaliter in statu suo perturbantur a potentiis, quae sunt ut quantitates materiae in illis contentae. Eorum igitur vires inertiae proportionales sunt his potentiis. Consequenter etiam materiae quantitatibus sunt proportionales." Euler, *Mechanica,* ed. Stäckel, p. 52.

mass (Definition 15)[9] Euler states explicitly that the matter (mass) of a body is not measured by its volume but by the force necessary to impart to it a given motion (acceleration). Here, then, is the earliest expression of the well-known formula "force equals mass times acceleration" and it serves as an accurate definition of mass.

The definition of mass as the ratio of force to acceleration gained wide acceptance, particularly within the French school of mathematical physicists. Most nineteenth-century treatises on *mécanique rationelle,* like those by Jean Marie Constant Duhamel[10] or by Henri Amé Résal,[11] begin their discussion with this definition. Paul Appell, for instance, in his classical *Traité de mécanique rationelle* writes: "The mass of a particle is the constant ratio that exists between the intensity of a constant force and the acceleration impressed by this force on the particle."[12]

However, the rise of modern foundational research, initiated in the middle of the nineteenth century by the development of non-Euclidean geometry and the study of its logical legitimacy, subjected the foundations of physics also to a close examination. The principles of Newtonian mechanics, in particular, became the object of a critical investigation by physicists, mathematicians, and philosophers like Saint-Venant, Reech, Andrade, Kirchhoff, Mach, Hertz, and Poincaré. The increasing influence of the new positivistic attitude toward the natural sciences led first, as we know,[13] to a severe criticism of the concept of force. What once, in Newtonian physics, played a central role was now regarded as an obscure metaphysical notion that has to be banished from

[9] "Id corpus plus materiae continere existimatur, non quod maius volumen occupat, sed ad quod dato modo movendum maior vis requiritur." *Ibid.,* p. 73.

[10] *Cours de mécanique* (Paris, 1845–1846), vol. 1, p. 93.

[11] *Traité de mécanique générale* (Paris, 1873–1881), vol. 1, p. 132.

[12] "La masse d'un point matériel est donc le rapport constant qui existe entre l'intensité d'une force constante et l'accélération qu'elle imprime au point." *Traité de mécanique rationelle* (Paris, 1893), vol. 1, p. 87.

[13] Max Jammer, *Concepts of force* (Harvard University Press, Cambridge, 1957), pp. 200–240.

science. It was claimed that kinematics, as the fusion of geometry with time, possessed logical and methodological priority over dynamics. The reduction of dynamic to kinematic conceptions was regarded as an important task for theoretical mechanics.

One of the earliest investigations carried out in this spirit was Barré de Saint-Venant's *Principes de mécanique fondés sur la cinématique,* published in 1851. In this work, as well as in numerous articles in the *Comptes rendus,* Saint-Venant rejects the traditional concept of a *quantitas materiae* as being void of any physical significance. "Mass," on the other hand, is for him a legitimate physical notion, definable in a "kinematic" way: the mass of a given body is the number of its parts which in mutual collision with an arbitrary body, taken as unity, impart to each other equal and opposite velocities. Or more accurately: "The mass of a body is the ratio of two numbers which express how often the body and a standard body contain parts which, if separated and then brought into mutual collision, two by two, communicate to each other opposite equal velocities." [14]

If two bodies, hurled against each other with equal velocities, also depart from each other with equal velocities, then their masses are equal. Saint-Venant now generalizes this definition. If v_1 and v_2 are the velocities before impact, $v_1 + \Delta v_1$, $v_2 + \Delta v_2$ after impact, and if m_1 and m_2 are the Newtonian masses, then, by the principle of conservation of linear momentum,

$$m_1 v_1 + m_2 v_2 = m_1(v_1 + \Delta v_1) + m_2(v_2 + \Delta v_2),$$

or

$$m_1 \Delta v_1 + m_2 \Delta v_2 = 0.$$

Consequently,

$$m_2 : m_1 = |\Delta v_1| : |\Delta v_2|.$$

Saint-Venant now uses the last equation as his point of departure

[14] "La masse d'un corps est le rapport de deux nombres exprimant combien de fois ce corps et un autre corps, choisi arbitrairement et constamment le même, contiennent de parties qui, étant séparées et heurtées deux à deux l'une contre l'autre, se communiquent, par le choc, des vitesses opposées égales." Barré de Saint-Venant, "Mémoire sur les sommes et les différences géometriques et sur leur usage pour simplifier la mécanique," *Comptes rendus 21,* 620 (1845).

for the definition of "equality of masses": two bodies have equal masses if their velocity increments after impact are equal. The original stipulation in Saint-Venant's definition, that is, the conceptual decomposition of the two bodies into elements of equal masses, was later omitted and the last formula itself was taken as the definition for the ratio of the masses of two bodies. Jules Andrade, in an article entitled *"Les idées directrices de la mécanique"*[15] and in his *Leçons de mécanique physique,* published in Paris in the same year (1898), regarded the formula

$$m_2 : m_1 = |\Delta v_1| : |\Delta v_2|$$

as the only unobjectionable definition of mass.[16]

In 1867 Ernst Mach, one of the most eloquent proponents of the rising antimetaphysical attitude toward science, suggested a new definition of mass. A five-page essay "On the definition of mass" in which he presented his ideas on this subject for the first time, was rejected by Poggendorf's *Annalen der Physik* but published a year later (1868) in *Carl's Repertorium der Experimentalphysik*.[17]

In accordance with the fundamental tenet of his doctrine that pure science as an abstract quantitative formulation of facts is not concerned with the elements of experience themselves but rather with the functional relations by which they are controlled, Mach rejected, like his predecessor Saint-Venant, the conception of *quantitas materiae* most categorically as well as even that of matter, as far as science is concerned. The notion of mass, however, designating a mathematical quantity that satisfies certain equations in theoretical physics, remains a concept of great con-

[15] *Revue philosophique de la France et de l'étranger 46,* 399–419 (1898).

[16] *Leçons de mécanique physique* (Paris, 1898), p. 53.

[17] Ernst Mach, "Über die Definition der Masse," *Carl's Repertorium der Experimentalphysik 4,* 355–359 (1868), reprinted in Ernst Mach, *Die Geschichte und die Wurzel des Satzes von der Erhaltung der Arbeit* (Prague, 1872), pp. 50–54. How Mach's approach to the concept of mass is related to the principle of conservation of energy, or, as Mach called it, "the principle of the impossibility of a *perpetuum mobile*" ("das Prinzip vom ausgeschlossenen *perpetuum mobile*") will be made clear later on.

venience for science.[18] Its methodological importance stems from the fact that, although its interpretation as a quantity of matter is an illegitimate transcendence beyond experience, it nevertheless is a specific constant characteristic of the physical body under discussion. This means, in particular, that its definition and determination must be independent of any particular physical hypothesis about what in Newtonian physics is called a field of force.

In order to obtain, therefore, a kinematic definition of force, Mach considers two particles *A* and *B* interacting with each other but otherwise unaffected by all the other particles in the universe. This assumption, he claims, is a legitimate extrapolation from experience. In his *Science of Mechanics*[19] he calls this statement the first experimental proposition. For the purpose of reference we quote the passage in full:

> *a.* First Experimental Proposition: Bodies set opposite each other induce in each other, under certain circumstances to be specified by experimental physics, contrary accelerations in the direction of their line of junction.
>
> *b.* Definition: The mass-ratio of any two bodies is the negative inverse ratio of the mutually induced accelerations of those bodies.
>
> *c.* Experimental Proposition: The mass-ratios of bodies are independent of the character of the physical states (of the bodies) that condition the mutual accelerations produced, be those states electrical, magnetic, or what not; and they remain, moreover, the same, whether they are mediately or immediately arrived at.

[18] "Wenn ich mich bemühe, alle metaphysischen Elemente aus den naturwissenschaftlichen Darstellungen zu beseitigen, so meine ich damit nicht, dass alle bildlichen Vorstellungen, wo dieselben nützlich sein können, und eben nur als Bilder aufgefasst werden, ebenfalls beseitigt werden sollen. Noch weniger ist aber eine antimetaphysische Kritik als gegen alle bisherigen wertvollen Grundlagen gerichtet anzusehen. Man kann z.B. ganz wohl gegen den metaphysischen Begriff "Materie" starke Bedenken haben, und hat doch nicht nötig den wertvollen Begriff "Masse" zu eliminieren, sondern kann denselben etwa in der Weise, wie ich es in der "Mechanik" getan habe, festhalten, gerade deshalb, weil man durchschaut hat, dass derselbe nichts als die Erfüllung einer wichtigen Gleichung bedeutet." Ernst Mach, *Die Prinzipien der Wärmelehre* (Leipzig, ed. 2, 1900), p. 363.

[19] Ernst Mach, *Die Mechanik in ihrer Entwicklung, historisch-kritisch dargestellt* (Leipzig, ed. 1, 1883); translated into English by T. J. McCormack under the title *The science of mechanics* (Open Court, La Salle, Ill., 1942), p. 304.

d. Experimental Proposition: The accelerations which any number of bodies $A,B,C, \ldots$ induce in a body K, are independent of each other.
e. Definition: Moving force is the product of the mass-value of a body into the acceleration induced in that body.

The following presentation of Mach's analysis of the concept of mass, although true in all details to the original text, will be given in a slightly modernized version.

Let $a_{A/B}$ denote the acceleration of particle A due to particle B and $a_{B/A}$ the acceleration of B due to A. Experience shows (1) that these accelerations are opposite in direction and (2) that their negative inverse ratio, denoted by $m_{A/B}$, is a positive numerical constant independent of the respective positions or motions of the particles:

$$m_{A/B} = -\frac{a_{B/A}}{a_{A/B}} = \text{a positive constant.} \tag{1}$$

If particle B is removed and replaced by a third particle C interacting with A, we obtain similarly

$$m_{A/C} = -\frac{a_{C/A}}{a_{A/C}} = \text{another positive constant.} \tag{2}$$

If only the particles C and B interact,

$$m_{C/B} = -\frac{a_{B/C}}{a_{C/B}} = \text{a third positive constant.} \tag{3}$$

Now, from experience we learn, as Mach proves by a thought experiment to be described presently, that the three mass ratios satisfy the following transitive relation:

$$m_{A/B} = m_{A/C} m_{C/B}. \tag{4}$$

Each mass ratio can consequently be represented as the ratio of two positive numbers:

$$m_{A/B} = \frac{m_A}{m_B}, \quad m_{A/C} = \frac{m_A}{m_C}, \quad m_{C/B} = \frac{m_C}{m_B}, \tag{5}$$

where the new constants m_A, m_B, m_C, any one of which may be chosen arbitrarily as unity, will be called the relative masses of the particles A, B, and C respectively. Equations (1), (2), (3) then read:

$$m_A a_{A/B} = -m_B a_{B/A}, \tag{1'}$$

$$m_A a_{A/C} = -m_C a_{C/A}, \tag{2'}$$

$$m_C a_{C/B} = -m_B a_{B/C}. \tag{3'}$$

These equations show that in the product of the relative mass of any particle and the acceleration induced by another particle the first factor is independent of the choice of the other particle. Equations (1′) and (2′), for instance, show that the mathematical description of the interaction of particle A either with B or with C associates with particle A the same relative mass m_A in both cases. Finally, if one of these particles is chosen as the standard particle with its relative mass taken as unity—if, say, $m_A = 1$—then the remaining relative masses—in our case m_B and m_C—are called simply the "masses" of the particles.

The transitivity of the mass ratios, in other words the validity of Eq. (4), can most easily be shown, according to Mach, in the particular case of equal masses, to which all other cases can be reduced. Equal masses are, of course, masses that induce mutually equal and opposite accelerations. The problem thus amounts to the following question, which may be called the "transitivity problem": Are two masses, say m_A and m_B, equal to each other if they are separately equal to a third mass, say m_C?

That the statement contained in Eq. (4) is logically equivalent to an affirmative answer on the "transitivity problem" follows from the following considerations: by Eq. (2′), $m_A = m_C$ implies

$$a_{A/C} = -a_{C/A}, \tag{2''}$$

and by Eq. (3′), $m_B = m_C$ implies

$$a_{C/B} = -a_{B/C}. \tag{3''}$$

Now, if also $m_A = m_B$, then by Eq. (1′)

$$a_{A/B} = -a_{B/A}. \tag{1''}$$

In other words, all six accelerations a are then antisymmetric in their indices. But then obviously the following equation holds:

$$-\frac{a_{B/A}}{a_{A/B}} = \left(-\frac{a_{C/A}}{a_{A/C}}\right)\left(-\frac{a_{B/C}}{a_{C/B}}\right), \tag{6}$$

which is the explicit formulation of Eq. (4).

In order to show that experience precludes the possibility of a negative answer to the "transitivity problem," Mach employs the following thought experiment. He imagines[20] three elastic bodies A, B, and C, movable on a fixed frictionless ring. Assuming that $m_{A/B} = 1$ (that is, $m_A = m_B$) and also $m_{B/C} = 1$ (that is, $m_B = m_C$), it follows from experience that $m_{A/C} = 1$. For if we impart a certain velocity to A, it will transfer by impact the same velocity to B, because of the first assumption, and B, in its turn, will transfer the same velocity to C, because of the second assumption. Now, if m_C were greater than m_A, A would receive an increased velocity so that the kinetic energy of the system would increase, in contrast to experience or in contrast to the "principle of the impossibility of a *perpetuum mobile*."[21] If m_A were smaller than m_C, a reversal of the direction of motion would lead to the same result.

Mach himself reports[22] that his ideas were at first ignored or bluntly rejected. With the publication of Kirchhoff's *Lectures on mechanics* (1874–1876), the declared objective of which was "to describe completely and in the simplest manner the motions which take place in nature,"[23] and owing to the great authority of Kirchhoff, Mach's conceptions became increasingly popular until, especially in the positivistic school of thought, they were hailed as a great advance in the development of mechanics.

One of the earliest objections to Mach's definition of mass was voiced with respect to the vagueness of the expression "under

[20] *Ibid.* (Leipzig, ed. 8, 1921), p. 214.
[21] See reference 17.
[22] Reference 20, p. 258.
[23] Gustav Kirchhoff, *Vorlesungen über Mechanik* (Leipzig, 1876).

certain circumstances to be specified by experimental physics," as Mach wrote in the First Experimental Proposition. Do these "circumstances" refer to the specification of a particular reference system? And worse, could it not be possible that the criterion for the choice of an admissible reference system presupposes the concept of mass?

That, in fact, the mass ratio $m_{A/B}$ does depend upon the system of reference S can easily be demonstrated. An observer whose reference system S' is moving with an acceleration a relative to S determines the mass ratio of the two particles A and B as

$$m'_{A/B} = -\frac{a'_{B/A}}{a'_{A/B}},$$

where $a'_{B/A} = a_{B/A} - a$ and $a'_{A/B} = a_{A/B} + a$.

His mass ratio is consequently related to $m_{A/B}$ by the equation

$$m'_{A/B} = m_{A/B} \frac{1 - (a/a_{B/A})}{1 + (a/a_{A/B})}. \tag{7}$$

Since the state of motion of the observer does not affect the condition of the "isolatedness" of the material system comprising the particles A and B, it is clear that every observer arrives, in general, at a different value for the mass ratio. In fact, for any arbitrarily given positive number y a transformation of the coordinate system can be established for any two given particles so that, with respect to the transformed system, y is the mass ratio of these particles. The new reference system is accelerated relatively to the original system with an acceleration

$$a = \frac{m_{A/B} - y}{(m_{A/B}/a_{B/A}) + (y/a_{A/B})}, \tag{8}$$

where $m_{A/B}$, $a_{A/B}$, and $a_{B/A}$ are the quantities, as defined above, with reference to the original system. On the other hand, it can be shown — and this is not a trivial result — that if in the reference system S two particles have equal masses, they have equal masses also in every other reference system S'. Putting $m_A = m_B$ (that is, $m_{A/B} = 1$) in Eq. (7), the antisymmetry of the accelerations $a_{A/B}$ and $a_{B/A}$ with reference to their indices yields $m_{A/B}' = 1$ or $m_A' = m_B'$. It is interesting to note that modern textbooks in mechanics which introduce the concept of mass more or less in accordance with Mach do not emphasize the important fact that the definition of mass depends on the choice of the reference system.

Generalizing these results, furthermore, it can be shown[24] that for any given dynamical system with masses $m_1, m_2, \ldots, m_n$ (relative to a reference system S) and for any arbitrarily given set of n positive scalars $m_1', m_2', \ldots, m_n'$ a reference system S' can be found relative to which these n scalars can be interpreted as the masses of the given particles.

The objection was also raised that Mach's definition implied the existence of forces whose exact nature was a matter of dispute. Was not the assumed "inducing" of accelerations an obscure, not to say mysterious, element in the whole train of thought, in open conflict with Mach's own principles? Volkmann thus questions the legitimacy of Mach's fundamental assumption that every two bodies, placed in opposition to each other, impart to each other opposite accelerations in the direction of their line of junction.[25] Wulf, in defense of Mach, suggests[26] that Mach's dynamical assumption be replaced by the apparently incontestable experimental fact that two centrally colliding bodies produce in each other opposite accelerations in the direction of the straight line through their centers of gravity.

Another difficulty to be overcome was Mach's assumption of a dynamically isolated system comprising only two bodies A and B. In practice, it was claimed, such isolated systems are rarely

[24] Paul Appell, "Sur la notion d'axes fixes et de mouvement absolu," *Comptes rendus 166*, 513–516 (1918).

[25] Paul Volkmann, "Über Newton's 'Philosophiae naturalis principia mathematica' und ihre Bedeutung für die Gegenwart," *Sitzungsberichte der physikalisch-ökonomischen Gesellschaft in Königsberg* (1898). See also *Beiblätter zu den Annalen der Physik und Chemie* (1898), pp. 917–918, and his *Einführung in das Studium der theoretischen Physik* (Leipzig, 1900).

[26] T. Wulf, "Zur Mach'schen Massendefinition," *Zeitschrift für physikalischen und chemischen Unterricht 12*, 205–208 (1899).

encountered. What, then, is the use of an operational definition, if its premise does not hold in nature? Astronomical binary systems (double stars), if located at considerable distances from other stars, could, of course, be taken as representations of such isolated systems in nature. The mathematical determination of the motions of their components, however, is a rather complicated task. It was natural to generalize Mach's procedure from a system of two particles to an arbitrary finite number of particles, so that, for instance, the solar system, which is perhaps the best-known approximately isolated system, can be regarded as a representation in nature. In connection with such a generalization, however, certain difficulties arise, as Pendse[27] and others pointed out. The acceleration a_A of particle A must first be resolved into components $a_{A/B}, a_{A/C}, \ldots, a_{A/N}$ in the directions of the lines joining A with $B, C, \ldots, N$; and the same must hold for a_B, a_C, and so on. Then it must be ascertained that the magnitude of each of these components depends exclusively on the positions of the two particles referred to in their respective indices. For the subsequent treatment it is more convenient to associate numerical indices with the individual particles. Thus particle A carries the index 1, particle B the index 2 and so on. Then the vectorial acceleration $\mathbf{a}_k$ of the kth particle ($k = 1, 2, \ldots, n$) can be represented by the following expression:

$$\mathbf{a}_k = \sum_{i=1}^{n} \alpha_{ki}\mathbf{u}_{ki} \quad (\alpha_{kk} = 0), \tag{9}$$

where $\mathbf{u}_{ki}$ is a unit vector in the direction from the kth particle to the ith particle and the α_{ki} are numerical coefficients. The $\mathbf{a}_k$ and $\mathbf{u}_{ki}$ are determinable by observation and measurement; the α_{ki} are unknowns to be determined.

Expressions (9) form a system of n vector equations in three-dimensional space or a system of $3n$ algebraic linear equations for

[27] C. G. Pendse, "A note on the definition and determination of mass in Newtonian mechanics," *Philosophical Magazine 24*, 1012–1022 (1937). See also C. G. Pendse, "On mass and force in Newtonian mechanics," *ibid. 29*, 477–484 (1940).

the $n(n-1)$ unknowns α_{ki}. The α_{ki} are therefore determinate only if $n(n-1) \leqslant 3n$, that is, if $n \leqslant 4$. Since, however, the α_{ki} determine the relative masses or mass ratios, it is clear that Mach's procedure breaks down for a system of five or more particles. For coplanar accelerations and motions, as exhibited approximately by the solar system, the procedure loses its validity for four, and for collinear accelerations for three particles.

In 1939 Pendse[28] generalized these conclusions on the basis of Newton's third law of motion as follows. Let there be given n particles with masses $m_k (k = 1, 2, \ldots, n)$ respectively and let $F_{ki}\mathbf{u}_{ki}$ be the force exerted on the kth particle by the ith particle in the direction of the unit vector $\mathbf{u}_{ki}$ from the kth to the ith particle. By Newton's third law, $F_{ki} = F_{ik}$. The equation of motion for the kth particle is

$$m_k \mathbf{a}_k = \sum_{\substack{i=1 \\ i \neq k}}^{n} F_{ki} \mathbf{u}_{ki}.$$

These are $3n$ distinct homogeneous linearly independent algebraic equations for $n + \frac{1}{2}n(n-1)$ unknowns (the n masses m_k and the $\frac{1}{2}n(n-1)$ forces F_{ki}). The solution is therefore indeterminate in general only if $n \geqslant 6$. If, moreover, the observations are performed at r different instants of time, the number of equations may rise to $3nr$ while the number of unknowns, because of the assumed constancy of the masses in time, is only $n + \frac{1}{2}n(n-1)r$. The equations thus will have an indeterminate solution for every r if $n \geqslant 8$. Finally, if the dynamical system of the n masses is dynamically isolated, that is, if the linear and angular momenta are conserved, the $6r$ homogeneous linear algebraic equations

$$\sum_{k=1}^{n} m_k \mathbf{a}_k(t) = 0,$$

$$\sum_{k=1}^{n} m_k [\mathbf{r}_k(t) \times \mathbf{a}_k(t)] = 0,$$

[28] C. G. Pendse, "A further note on the definition and determination of mass in Newtonian mechanics," *Philosophical Magazine 27,* 51–61 (1939).

observed at r different instants of time t, determine the $n - 1$ mass ratios uniquely (apart from a coefficient of proportionality) if they are independent, provided that $r > \frac{1}{6}(n - 1)$. Thus, under certain conditions, the masses of a dynamical system are determinable, provided a sufficiently large number of observations is made.

If we ignore such subtle difficulties as outlined so far, we may rightly regard Mach's definition of mass as an acceptable operational definition of a theoretical construct. By establishing a rule of correspondence between the theoretical construct "mass" and a series of clearly defined operations (measurements of acceleration), the theorems of theoretical mechanics, as far as the concept of mass is concerned, become a meaningful set of propositions. Among the earliest philosophers of science who realized the importance of Mach's conception of mass, William Kingdon Clifford and Karl Pearson must be mentioned. In *The common sense of the exact sciences,*[29] published in 1885 in the International Scientific Series (London), Clifford defines the concept of mass wholly in the Machian manner. Pearson, in his at the time very popular *Grammar of science,* published originally in 1889 in "The Contemporary Science Series" (London), speaks of the "scientific conception of mass" which is essentially Mach's approach.[30] Dühring, on the other hand, in his *Critical history of the general principles of mechanics,*[31] published five years after Mach's publication on "mass," still identifies "mass" with "quantity of matter,"[32] basing his reasoning on what may be called the additivity of mass. Two bodies, he argues, that are alike in all respects, have, if combined in any way whatever, twice as much mass as a compound body as each component had before. This property, he continues, assures a methodological

[29] Reprinted by Dover, New York, 1955. See chap. 5, sec. 11, "Of mass and force," especially p. 241.

[30] See chap. 9, sec. 8 of the 1892 edition (London).

[31] E. Dühring, *Kritische Geschichte der allgemeinen Principien der Mechanik* (Berlin, 1873).

[32] "Die Masse oder, mit anderen Worten, die Menge der Materie." *Ibid.,* p. 199.

foundation and logical autonomy for the concept of mass independent of the science of mechanics. Although the quantitative determination is a matter of dynamics, he adds, the very notion of mass is prior to such determination, and without such an antecedent conception the procedure leading to the numerical determination would have been inconceivable.[33] In view of such considerations it is not surprising that in 1909 a prominent philosophical dictionary still defined the concept of mass as quantity of matter.[34] Another more recent example (1949) is Abram Fedorovich Ioffé's book *Basic concepts in contemporary physics,* translated by the Soviet National Science Foundation, Moscow; its proclaimed purpose is the vindication of the philosophical tenets of dialectical materialism through a study of contemporary physics. Because of the important role that the notion of mass plays in the materialistic philosophy of Lenin, it is of special interest to see how the late director of the Leningrad Physico-Agronomy Institute conceived the idea of mass. "Mass," as defined by Ioffé, is "the measure of quantity of matter."

In general, however, it must be admitted that the textbooks of physics that were written or published since the turn of the century only rarely restate the vague conception of mass as quantity of matter. Among more than 140 treatises and textbooks on physics, examined in this respect in 1918 by Edward V. Huntington,[35] only a single one defined mass as merely "quantity of matter."[36] The majority of the textbooks examined introduce the concept of mass as the quotient of force and acceleration or by the operation of weighing by a balance. About 10 out of these

[33] "Zwei in allen Beziehungen gleiche Körper . . . in irgendeiner Form vereinigt . . . [haben] die doppelte Masse. Hindurch ist klar, dass der Begriff der Masse . . . etwas durchaus Selbständiges ist und nicht etwa aus der Mechanik selbst concipiert wird." *Ibid.,* p. 201.

[34] Eli Blanc, *Dictionnaire de philosophie* (Paris, 1909), p. 804: "Ce qu'on peut dire de plus clair peut-être c'est que la masse d'un corps est la quantité de matière de ce corps."

[35] "Bibliographical note on the use of the word mass in current textbooks," *American Mathematical Monthly 25,* 1–15 (1918).

[36] Alexander Ziwet, *Theoretical mechanics* (New York, 1906).

140 textbooks follow Mach's procedure more or less closely. An unpublished statistical study of textbooks in physics and treatises on mechanics (or dynamics) carried out by the present author for the period from 1920 to 1960 shows clearly that the percentage of books defining "mass" by mutual accelerations increased considerably.

The logical and methodological aspects of Mach's definition were further elaborated especially by logical empiricists such as Rudolf Carnap[37] and Philipp Frank.[38] Carnap showed how the relation "equality of masses" (corresponding to equal accelerations) is a transitive and symmetric relation and "inequality of masses" is a transitive and asymmetric relation thus complying with the rules of what he terms "topological definitions" which make it possible to associate unambiguously a numerical value with the property under discussion.

A different approach to a precise operational definition of mass was chosen by those theorists who, following Euler in his conception of mass as the ratio of force to acceleration, did not object to the use of "force" as a physically significant concept. One of the most elaborate expositions along these lines was given by Maxwell in his *Matter and motion.*[39] For the definition of equal masses Maxwell assumes "that it is possible to cause the force with which one body acts on another to be of the same intensity on different occasions." Thus, if the same rubber band or the same elastic spring, for instance, stretched to the same elongation, imparts at the end of a unit of time the same velocities to two bodies, the masses of these bodies are defined as equal. "This is the only definition of equal masses which can be admitted in dynamics, and it is applicable to all material bodies, whatever they may be made of."[40] By methods of comparison,

[37] Rudolf Carnap, *Physikalische Begriffsbildung* (Karlsruhe, 1926), pp. 42–43.

[38] Philipp Frank, *Philosophy of science,* pp. 111–116, 131–133.

[39] James Clerk Maxwell, *Matter and motion, with notes and appendices by Sir Joseph Larmor* (Dover, New York, no date). The first edition of the book appeared in 1876 in "Manuals of Elementary Science."

[40] *Ibid.,* p. 34.

based on the additivity of mass, and by calibration, masses of different bodies can be compared; by defining an arbitrary unit of mass, a scale for mass determination can be established. The characteristic feature of Maxwell's method is the fact that it assigns priority to the concept of force over that of mass, a step which, as we have seen, Mach's definition just tried to avoid.

An ardent proponent of the priority of "force" over "mass" was Alois Höfler. In a commentary to Kant's *Metaphysical foundations of natural science* and in an article published by the Viennese Philosophical Society[41] Höfler bases the definition and determination of mass on that of force or tension. "A mass of one gram is a body which subjected to a tension of one dyne acquires an acceleration of one $cm\ sec^{-2}$." [42] Höfler justifies his definition of mass with the remark: "The tonomonic quantity 'dyne' precedes logically the notion of 'one gram mass.' " [43]

Höfler, who claimed that psychologic-didactic considerations lend support to his ideas, was not alone in his stand. A few years earlier, Clémentitch de Engelmeyer had published in the *Revue philosophique de la France et de l'étranger* an interesting investigation concerning the sensory origin of scientific concepts employed in mechanics in which he emphasized "that our daily experience prepares us much better for a comprehension of the notion of force than of mass . . . Mass thus should resume the role of a derived quantity." [44] Instead of the conventional cgs system which after Gauss is usually called the "absolute sys-

[41] Alois Höfler, "Studien zur gegenwärtigen Philosophie der Mechanik," *Veröffentlichungen der philosophischen Gesellschaft an der Universität zu Wien*, vol. 3 (b), 1900.

[42] Alois Höfler, ed., *Kants Metaphysische Anfangsgründe der Naturwissenschaft* (Leipzig, 1900), p. 76.

[43] See also Alois Höfler, "Einige Bemerkungen über das C.S.G.-System im Unterricht," *Zeitschrift für den physikalischen und chemischen Unterricht 11*, 79 (1898).

[44] Clémentitch de Engelmeyer, "Sur l'origine sensorielle des notions mécaniques," *Revue philosophique de la France et de l'étranger 39*, 511–517 (1895): "Les études psycho-physiologiques démontraient que notre expérience de chaque jour nous prépare mieux à comprendre la notion de la force que celle de la masse. Pour ma part je crois qu'il en est ainsi. La masse alors rentrerait parmi les autres grandeurs dérivées" (p. 517).

tem,"[45] Engelmeyer proposes a "didactic system" based on length (L), time (T) and force (F) as the fundamental dimensions, so that mass becomes a derived quantity with the dimension FT^2L^{-1}.

For the sake of historical accuracy and completeness a third approach to a definition of "mass" has to be mentioned, a method to which allusion has already been made in connection with the various interpretations of Newton's definition of mass. It is based on the purely conjectural assumption that the ultimate particles of all substances are fundamentally the same. On this assumption the mass of a body is defined simply as the number of material particles. *Johnson's cyclopedia*[46] and *La grande encyclopédie*[47] define "mass" as "the number of identical particles."

A prima facie similar definition of mass was advanced by Heinrich Hertz in his *Principles of mechanics.* "The number of material particles in any space, compared with the number of material particles in some chosen space at a fixed time, is called the mass contained in the first space."[48] Hertz, however, qualifies his definition by the following remark: "We may and shall consider the number of material particles in the space chosen for comparison to be infinitely great. The mass of the separate material particles will therefore, by the definition, be infinitely small. The mass in any given space may therefore have any rational or irrational value." This somewhat paradoxical statement — how can a ratio of numbers of particles be irrational? — may find its explanation in Hertz's concept of "material particle."

[45] K. F. Gauss, *Intensitas vis magneticae ad mensuram absolutam revocata* (Göttinger Abhandlungen, 1832); reprinted in *Werke* (Leipzig, 1863–1903), vol. 5, pp. 81–118. Gauss actually adopted length, mass, and acceleration (*vis acceleratrix*) as fundamental dimensions.

[46] *A. J. Johnson's new universal cyclopedia* (New York, 1893–1895), vol. 2, p. 875, "Dynamics."

[47] Paris, 1886–1902, vol. 23, p. 371, "Mass: astronomy."

[48] Heinrich Hertz, *Die Prinzipien der Mechanik in neuem Zusammenhang dargestellt* (Leipzig, 1894); trans. into English by D. E. Jones and J. T. Walley (London, 1899); reprinted, with a new introduction by R. S. Cohen (Dover, New York, 1956) under the title *The principles of mechanics presented in a new form,* p. 46.

For this is not to be understood as an atomic unit or indivisible element, but merely as a concept of internal intuition by which "we associate without ambiguity a given point in space at a given time with a given point in space at any other time."[49] In other words, owing to its indestructibility and unchangeability it is the carrier of identity and as such from the mathematical point of view on the same footing as a "point" in Euclidean geometry. As soon, however, as we conceive masses as symbols for objects of external experience, Hertz argues, the previous definition has to be supplemented by procedures that determine the relations between sensory perceptions. Such a procedure, consistent with the previous theoretical definition, is according to Hertz the determination of mass by weighing: "The mass of bodies that we can handle is determined by weighing."[50]

Hertz's operational definition of mass thus belongs essentially to that class of definitions which define mass by weight. As we have mentioned above, numerous textbooks in the beginning of this century define equality of mass and mass in general by the use of a balance (beam balance). To quote a typical example, Ganot[51] defines equality of masses as follows: "Two bodies are said to have equal masses when, if placed in a perfect balance in vacuo, they counterpoise each other." Picard's definition of mass by means of the deformation of an elastic beam may be reckoned to the same category of definitions.[52]

Defining "mass" by "weight" is perhaps from the practical point of view a defensible method, since it undoubtedly describes the most efficient way to determine the masses of ordinary physical objects. From the didactic point of view, however, it easily leads to the confusion of mass and weight. From the methodological point of view it employs a purely accidental aspect in classical physics, the proportionality of gravitational

[49] *Ibid.,* p. 45.
[50] *Ibid.,* p. 140.
[51] *Ganot's Physics,* trans. by Atkinson and Reinold (Longmans, New York, 1906, ed. 17), p. 15.
[52] Emile Picard, *Quelques réflexions sur la mécanique* (Paris, 1902), p. 43.

and inertial mass, for the definition of the latter. If, for instance, Newton's law of gravitation read $F = GM^n m^n / r^2$, where M and m are the masses of the mutually attracting bodies, r the distance between them, G the constant of gravitation, and n a number different from unity, then clearly the ratio of the weights of two objects would be equal to the nth power of the ratio of their masses.

Definitions of this category become certainly less objectionable if they admit explicitly that they do not express the characteristic feature of the definiendum but are only definitions by abstraction. Thus, for instance, Harold Jeffreys states explicitly: "We classify together bodies that counterbalance the same body, and abstract the property of mass. If then a body counterbalances the same body as is counterbalanced by n of our standard weights, we say that its mass is n in terms of these weights." [53]

A rather sophisticated derivation of the concept of mass, also based on the notion of force, was given in 1911 by Anton Lampa.[54] In the first place, Lampa declares, experiments with the Atwood machine show that a constant force gives rise to a uniformly accelerated motion. If F denotes the accelerating force and a the acceleration of the body B, a functional relation between F, a, and a characteristic property of B is set up:

$$F = f(B, a). \tag{10}$$

In order to determine the structure of the function f, recourse is made to experiments with the well-known centrifugal-force apparatus in which the centrifugal force exerted on a body, sliding on a smooth rotating shaft, is counterbalanced by a weight attached symmetrically to the axis of rotation and connected by a cord to the revolving body. If the number of revolutions per second is doubled, trebled, and so on, the centripetal acceleration, as shown in kinematics, increases by a factor of 4, 9, and

[53] Harold Jeffreys, *Scientific inference* (Macmillan, New York and London, 1931), p. 89.

[54] Anton Lampa, "Eine Ableitung des Massenbegriffs," *Lotos, Naturwissenschaftliche Zeitschrift* (Prague) *59*, 303–312 (1911).

so on, provided the radial distance r of the body from the axis of rotation is kept constant. Experiments further show that the body B is balanced by four, nine, and so on, times the original weight. The centrifugal force is thus proportional to the centripetal acceleration and Eq. (10) must conform to this condition:

$$nF = f(B, na), \tag{11}$$

which can be satisfied only if

$$F = a\varphi(B), \tag{12}$$

where $\varphi(B)$ is a characteristic property of B and determines the acceleration under the action of F. Referring now to experiments with two sliding metal bodies B_1 and B_2 of weights G_1 and G_2, connected by a cord and attached to the same horizontal rotating shaft, the function φ in Eq. (12) can be determined. In the case of equilibrium, the centrifugal forces balance each other, independently of the number of rotations per second:

$$F_1 = a_1\varphi(B_1) = F_2 = a_2\varphi(B_2). \tag{13}$$

If the radial distances of B_1 and B_2 from the axis of rotation are respectively R_1 and R_2 and the angular velocity is ω, then $a_1 = R_1\omega^2$, $a_2 = R_2\omega^2$, so that by Eq. (12)

$$R_1\varphi(B_1) = R_2\varphi(B_2). \tag{14}$$

Comparison of the radii R_1 and R_2 with the weights G_1 and G_2 of the bodies B_1 and B_2 shows that $R_1G_1 = R_2G_2$. Therefore, in virtue of Eq. (14),

$$\frac{\varphi(B_1)}{G_1} = \frac{\varphi(B_2)}{G_2}. \tag{15}$$

In other words,

$$\varphi(B) = CG, \tag{16}$$

where C is a constant independent of the particular choice of the body B. Equation (12) now shows that

$$F = aCG, \tag{17}$$

which makes it possible to compute the acceleration a of a body of weight G, acted on by a given force F, provided the constant C is known. Finally, in order to determine C, Eq. (8) is applied to the case of free fall, in which the accelerating force is the weight G itself of B:

$$G = gCG, \tag{18}$$

where g is the acceleration of free fall. The last equation yields

$$C = 1/g, \tag{19}$$

and consequently, the characteristic property $\varphi(B)$ of body B turns out to be G/g, which is called "the mass" of body B. Needless to say, Lampa's definition of mass, interesting as it is from a conceptual point of view, is too complicated to serve as an introduction to this important and fundamental concept.

If for certain theorists at the turn of the century "force" had priority over "mass," for Wilhelm Ostwald it was the notion of energy that was the foremost fundamental conception in physical science. Mass had thus to be defined in terms of energy. In his *Lectures on natural philosophy*[55] Ostwald raises the question whether the energy of motion of a physical object, apart from its dependence on velocity, is a function of any further variables. That the answer is affirmative is demonstrated by a simple experiment: a piece of cork and a stone are hurled against our body with the same velocity. We experience a much stronger blow in the case of the stone than of the cork. We also know from experience that it requires more work (energy) to impart a certain velocity to the stone. "This special property on which the energy of a moving body depends, apart from its velocity, is called mass."[56] Thus mass, for Ostwald, is merely a capacity for kinetic energy, just as specific heat is a capacity for thermal

[55] Wilhelm Ostwald, *Vorlesungen über Naturphilosophie* (Leipzig, ed. 2, 1902), p. 184.

[56] "Diese besondere Eigenschaft, von der die Energie eines bewegten Körpers ausser seiner Geschwindigkeit abhängt, nennt man Masse." *Ibid.*, p. 185.

energy. Mathematically speaking, mass is defined and determined by the expression

$$\frac{E}{\frac{1}{2}v^2},$$

where E is the kinetic energy and v is the velocity of the moving object.

In a lecture delivered in 1895 at a scientific congress in Lübeck, Ostwald emphasized the consistency between his concept of mass and his general "energetics," according to which ponderable or tangible matter is but a collocation of energy. Ostwald's definition of mass was soon generalized beyond mechanical phenomena for electromagnetic processes by von Türin.[57] Ostwald's conception of the mass of a physical object in terms of energy and velocity, of its volume in terms of compressibility, and its shape in terms of elasticity, is one of the final stages in a development that began in John Locke's sensationalistic philosophy and eventually put an end to the substantial conception of matter. What we sense is no longer the dubious and in itself entirely indefinite "matter," which the early proponents of classical mechanics thought to have brought within the compass of exact quantitative determination by their concept of *quantitas materiae* or *vis inertiae*. The passive and indifferent substratum of properties is now set aside. The object is only what it appears to be: a source of activities, of ways to affect our senses. This trend toward a desubstantialization of matter is to some extent implicit already in Mach's conception of mass. It is certainly one of the basic tenets of Ostwald's conception of mass within the framework of his natural philosophy.

Reviewing the situation at the beginning of the present century, we see ourselves confronted with a confusing choice of different definitions of mass. The perplexity of the situation was fully realized by contemporary scholars and scientists, and the

[57] V. von Türin, "Über die Grundsätze und Hauptbegriffe der Mechanik," *Annalen der Naturphilosophie* 5, 378–394 (1906).

problem of mass was a topic of much discussion in professional journals and a matter of much debate in scientific conferences. The Italian Physical Society, for instance, thought it necessary to appeal to the outstanding theorists of mechanics in that country — and mechanics, like geometry, had a stronghold at that time in Italy — in quest for a clarification of the situation. As a result of this appeal G. Vanni, L. Silla, C. Goretti, E. Alessandrini, M. Ascoli, F. Bonetti, D. Mazzotto, G. Castelnuovo, and others participated in 1907 in a discussion on the concept of mass and its presentation in the elementary instruction of mechanics. The minutes of this debate appeared under the title "Il concetto di massa nell' insegnamento elementare della meccanica" in the *Nuovo Cimento* of that year.[58] A unanimous agreement on how to introduce the concept of mass in courses on mechanics has not been reached and, in fact, remains a question of some debate today.

[58] "Discussione fatta in seno alla Societa Italiana di Fisica," *Nuovo cimento* [5] *14*, 80–124 (1907).

CHAPTER 9

THE CONCEPT OF MASS IN AXIOMATIZED MECHANICS

The modern axiomatization of mechanics, like the axiomatizations of numerous other branches of science (except purely mathematical disciplines like algebra or geometry), was not the outcome of needs intrinsic to the specialized research in the subject; rather it was carried out to satisfy a general philosophic-aesthetic longing for a finished conceptual structure characterized by a high degree of mathematico-logical rigor. In comparison with axiomatizations of biology,[1] of music,[2] or of psychology,[3] for instance, an axiomatization of mechanics seems to be a relatively easy task, since, with the exception of a few dynamic concepts and the notion of time, geometric considerations play a predominant part, and geometry is one of the most rigorously axiomatized systems in scientific thought.

Now, one of these dynamic concepts that distinguish mechanics proper from geometry is just the concept of mass. It may therefore be expected that axiomatic investigations on mechanics contribute perhaps to the logical and methodological clarification of our concept.

[1] J. H. Woodger, *The axiomatic method in biology* (Cambridge University Press, New York, 1937).

[2] Susanne K. Langer, "A set of postulates for the logical structure of music," *Monist 39*, 561–570 (1929).

[3] J. H. Woodger, "The formalization of a psychological theory," *Erkenntnis 7*, 195–198 (1937).

As we have seen in the last chapter, various physical laws or hypotheses can be chosen as a basis for the definition of mass. If "mass" is taken as a derived notion within the framework of the deductive system, these laws or hypotheses have to be accounted for and have to be included in the formalization. In order to avoid the difficulties involved in such a procedure, it is natural to adopt the notion of mass as a primitive concept. In fact, most axiomatizations of mechanics[4] adopted the notion of mass as a primitive concept in addition to the undefined notions of position, time, and particle (or set of particles). Such an approach is wholly adequate for the elucidation of the formal and analytic aspects of the system. If, however, the "metaxiomatic" requirement of a correspondence between primitive notions (at the formal axiomatized level) and observables (at the operational, empirical level) is stipulated — a requirement that naturally has no analogue in the axiomatization of purely mathematical theories — then the concept of mass becomes necessarily a definiendum in the formalized system.

The difficulties encountered in taking "mass" as a definable notion apparently can be reduced to a minimum, if the physical law or hypothesis underlying the definition is of maximal simplicity. Hans Hermes, in his attempt to axiomatize mechanics in accordance with the above-mentioned requirement, saw in the mechanism of inelastic collision (subject to conservation of momentum) the simplest physical law lending itself to a formalized definition of mass.[5] The following is a condensed and simplified version of a paper read by Hermes[6] at the international

[4] For example, J. C. C. McKinsey, A. C. Sugar, and P. Suppes, "Axiomatic foundations of classical particle mechanics," *Journal of Rational Mechanics and Analysis 2,* 253–272 (1953). See also H. Rubin and P. Suppes, "Axioms of relativistic particle mechanics," in "Transformations of systems of relativistic particle mechanics," *Pacific Journal of Mathematics 4,* 563–601 (1954).

[5] Hans Hermes, *Eine Axiomatisierung der allgemeinen Mechanik* (Forschungen zur Logik und zur Grundlegung der exacten Wissenschaften, Heft 3; Leipzig, 1938).

[6] Hans Hermes, "Zur Axiomatisierung der Mechanik," in *The axiomatic method, Proceedings of an International Symposium,* ed. L. Henkin, P. Suppes, and A. Tarski (North Holland, Amsterdam, 1959), pp. 282–290.

symposium on the axiomatic method held at the University of California, December 26, 1957 to January 4, 1958. Temporal cross sections through the world line of an unextended particle, called a "momentary mass point," will be denoted by x, y, . . . Two momentary mass points x, y, belonging to the same (physical) particle — in other words, two temporal cross sections of the world line of the same particle — are called "genidentical," a relation to be symbolized by Gxy. If S is an inertial reference system, $CStxy$ expresses the fact that the momentary mass points x and y collide inelastically at the time t with respect to an S in which their common velocity after impact equals zero. Finally, Vel_Svtx denotes the velocity of x with respect to S immediately before the collision at time t. Hermes's definition of the mass ratio of the two particles is then given by the following expression:

$$\textit{Definition:}\ \text{Mass}\ \alpha x x_0 =_{\text{Df}} \bigvee_S \bigvee_t \bigvee_y \bigvee_{y_0} \bigvee_v \bigvee_{v_0} (Gxy \wedge Gx_0y_0 \wedge CStyy_0 \wedge Vel\_Svty \wedge Vel\_Sv_0ty_0 \wedge \alpha|v| = |v_0|) \vee (Gxx_0 \wedge \alpha = 1).$$

In this definition, $\bigvee$ is the existential quantifier, $\vee$ the symbol for logical disjunction, and $\wedge$ that for logical conjunction. In nontechnical language the definition says: "The mass of x is α times that of x_0" means "There exist a system S, an instant t, and momentary mass points y, y_0, genidentical with x, x_0 respectively, whose velocities (immediately before impact) v and v_0 are in the ratio $|v_0|:|v| = \alpha$. If x and x_0 are themselves genidentical, their mass ratio is unity."

In order to fully understand this definition, let us recall that immediately after the inelastic collision the colliding particles move with a common velocity. Thus an inertial system can be found relative to which this common velocity equals zero. For this S the classical conservation law states (in conventional symbols):

$$m_1v_1 + m_2v_2 = 0,$$

or

$$m_1 : m_2 = |v_2| : |v_1|,$$

where v_1 and v_2 are the velocities before impact. Thus, if $|v_0| = \alpha|v|$, the particle with the momentary mass point x has a mass α times as great as that whose momentary mass point is x_0. This, in simple words, is the contents of the above-mentioned definition. In order that this newly defined relation between x and x_0 be really a ratio of two numbers, an additional axiom has to be introduced:

Axiom: $\textit{Mass}\ \alpha yz \wedge \textit{Mass}\ \beta zx \wedge \textit{Mass}\ \gamma xy \longrightarrow \alpha\beta\gamma = 1$,
where $\longrightarrow$ denotes the implication symbol. On the basis of this axiom and other postulates some important theorems can be proved:

Theorem: $C\ Stx_1x_2 \longrightarrow C\ Stx_2x_1$;
Theorem: $\textit{Mass}\ \alpha xx_0 \longrightarrow \alpha \neq 0$
Theorem: $\textit{Mass}\ \alpha xx_0 \longrightarrow \textit{Mass}\ \frac{1}{\alpha}\ x_0x$;
Theorem: $\textit{Mass}\ \alpha xx_0 \wedge Gxy \wedge Gx_0y_0 \longrightarrow \alpha = \beta$.

The meaning and physical significance of these theorems are easily understood. Thus, the last theorem, for instance, states that the choice of momentary mass points is irrelevant for the numerical mass ratio of the particles that are represented by these momentary mass points.

Hermes's approach, like that of Mach, defines the concept of mass, of course, only in so far as it specifies a procedure for associating positive number, the mass ratio, with a given set of two particles. It is important for the logical comprehension of this process but does not supply any additional insight into the physical significance of the concept. One may, of course, argue that this is all that matters. According to this view, physical science is but a list of stipulations or specifications by which numbers are associated with certain operations or observations and the functional dependence among these numbers is the object of investigation. The physical significance of a concept is thus reduced to the mathematical relations into which the number, associated with the concept, enters with the numbers relating to other concepts or phenomena. The analytical apparatus to be employed for revealing the significance of the concept has thus to be based on the methods of statistical inference.

The earliest allusion to such an approach as far as the concept of mass is concerned was made in 1939 by C. G. Pendse. While considering an isolated system composed of n masses m_1, m_2, . . . , m_n, which in Newtonian physics satisfies the equation

$$\sum_{k=1}^{n} m_k \ddot{\mathbf{r}}_k = 0, \tag{1}$$

Pendse raises the problem: Is one "to assume the property of inertia, i.e. the existence of n positive numbers, each number being associated with one particle, one of the numbers being chosen arbitrarily? . . . If he makes these assumptions, his work reduces to that of computing the ratios of the masses of the particles and the interactions between them: his problem is one of computation."[7] While for Pendse the m_k's in Eq. (1) were, however, still the masses of the n particles as previously defined according to Mach's definition, Herbert A. Simon uses Eq. (1) and the corresponding equation for the conservation of angular momentum,

$$\sum_{k=1}^{n} m_k(\mathbf{r}_k \times \ddot{\mathbf{r}}_k) = 0, \tag{2}$$

as a definition of the m_k's on the basis of statistical inference in analogy to methods used in econometrics.[8]

Usually, the object of any investigation in dynamics is the computation of the motions or trajectories of the particles under discussion. Simon[9] reverses the problem and tries to infer from

[7] C. G. Pendse, "A further note on the definition and determination of mass in Newtonian mechanics," *Philosophical Magazine 27,* 55 (1939).

[8] Cf. C. Koopmans, ed., *Statistical inference in dynamic economic models* (Wiley, New York, Chapman & Hall, London, 1950), especially part 1, "Simultaneous equation systems," pp. 1–265. Cf. also A. Wald, *Selected papers in statistics and probability* (McGraw-Hill, New York, Toronto, London, 1955), "Note on the identification of economic relations," pp. 569–575.

[9] Herbert A. Simon, "The axioms of Newtonian mechanics," *Philosophical Magazine 38,* 888–905 (1947).

the known or observed motions the physical significance of the mathematical entities by means of which the motions are calculated. In view of the above-mentioned metaxiomatic requirement which Simon fully accepts, concepts like force or torque, as being nonobservables, have to be definable notions.[10] Consequently, Eqs. (1) and (2) cannot be taken as deductions from a preceding law of force, but have to be taken as definitions for "isolated motions," provided the scalar quantities m_k are positive quantities. Taking into account the possibility of only positive masses, Simon's introduction of the concept of mass begins with the following preliminary definitions:

Definition 1: A motion, (π), is a finite point set, $\{P_i\}$ $(i = 1, 2, \ldots, n)$, on which is defined, for $t_1 \leqslant t \leqslant t_2$, a regular vector function $\mathbf{r}_i(t) = \mathbf{r}_i[x_i(t), y_i(t), z_i(t)]$. Here x, y, z, t are real numbers and x, y, z are single-valued functions of t.

Definition 2: The vector $\mathbf{r}_i(t)$ is called the position of P_i at t in the motion (π).

Definition 3: If there exists a set of scalars $\{m_i\}$, $m_i > 0$ $(i = 1, 2, \ldots, n)$, constant with respect to t, such that (a) $\sum_{i=1}^{n} m_i \ddot{\mathbf{r}}_i(t) \equiv 0$, (b) $\sum_{i=1}^{n} m_i[\mathbf{r}_i(t) \times \ddot{\mathbf{r}}_i(t)] \equiv 0$, both identically in t, then the set $\{P_i\}$, together with the associated vectors $\{\mathbf{r}_i(t)\}$ and scalars $\{m_i\}$ is called a proper isolated motion, $[[\pi]]$.

Definition 4: The element P_i, together with the functions $\mathbf{r}_i$ and m_i defined on it, is called a mass point or particle in $[[\pi]]$.

Thus a particle in Simon's axiomatic system is defined as a point (which is a primitive notion in this system) associated with a trajectory $\mathbf{r}_i$ and a positive scalar m_i. The concept of mass is introduced by the following definition:

Definition 5: The positive scalar m_i is called the mass of P_i in $[[\pi]]$.

The m_i's of Definition 3 are not necessarily uniquely determined, as has been shown already by Pendse[11] by the example

[10] Herbert A. Simon, "The axiomatization of classical mechanics," *Philosophy of Science 21*, 340–343 (1954).

[11] See reference 7.

of a uniformly rotating regular polygon of n sides whose vertices are the loci of particles with masses $m_1, m_2, \ldots, m_n$. A necessary and sufficient condition for the uniqueness of the masses m_i — within a factor of proportionality, of course — is given by the following theorem:

Theorem: The masses are uniquely determined if and only if there exists no proper subset (π') of the motion (π) on which an isolated motion can be defined.

A motion is called disjunct if $\ddot{\mathbf{r}}_i(t) \equiv 0$. The motion of the fixed stars, to a good approximation, is disjunct. It follows from the theorem just quoted that the masses of a disjunct motion are not uniquely determined. The problem of how to select an appropriate reference system relative to which the masses, if uniquely determined, have their usual physical significance can be solved, according to Simon, only by an additional physical hypothesis:

Physical hypothesis: There exists a Galilean class of reference systems with respect to which the motion of the fixed stars (excepting double stars) is a disjunct motion.

Masses determined with respect to such a reference system are called Newtonian inertial masses.

According to the present formalization of mechanics, "masses" are defined by their mathematical determination as numerical coefficients in certain definitory equations which characterize certain classes of motions or systems. Mach's conception of mass as a mathematical quantity "that merely satisfies an important equation"[12] is fully realized in Simon's presentation of the concept.

However, certain objections can be raised against the logical legitimacy of Simon's method of defining "mass," in view of modern definitions of definability. Since Tarski's important investigations on the parallelism between "axioms," "theorems," and "proofs" on the one hand, and "primitive notions," "definable

[12] ". . . nichts als die Erfüllung einer wichtigen Gleichung . . ." Ernst Mach, *Die Prinzipien der Wärmelehre* (Leipzig, ed. 2, 1900).

terms," and "definitions" on the other,[13] the question whether the primitive notions of a given deductive theory are mutually independent became the object of methodological research—just as the problem of the logical independence of axioms (as well as their completeness and consistency) became an important topic of foundational research after the publication of David Hilbert's *Foundations of geometry*.[14]

Tarski defines an extralogical constant a as definable, with respect to a set X of sentences (for example, a deductive theory), in terms $b_1, b_2, \ldots$ of the set B (on the basis of X), if a and the terms of B occur in the sentences of the set X, and if at least one sentence of the following structure is derivable from the sentences of X:

$$(x) : x = a \cdot \equiv \cdot \phi(x; b_1, b_2, \ldots),$$

where ϕ denotes any sentential function which contains x as the only real variable and in which the only extralogical constants (primitive or defined) are the $b_1, b_2, \ldots$ of B. In nontechnical language the contents of the above formula may be described as follows: "ϕ is a definition of a if, for every x, x satisfies ϕ if and only if x is identical with a." On the basis of this definition of definability the following theorem can be proved:

Theorem: The term a is definable, in the sense of the previous definition, if and only if the formula

$$(x_1, x_2) : \phi(x_1 : b_1, b_2, \ldots) \cdot \phi(x_2 : b_1, b_2, \ldots) \rightarrow x_1 = x_2$$

is logically demonstrable.[15]

[13] Alfred Tarski, "Z badań metodologicznych nad definiowalnością terminów," *Przegląd filozoficzny* 37, 438–460 (1934); reprinted in condensed form under the title "Einige methodologische Untersuchungen über die Definierbarkeit der Begriffe," *Erkenntnis* 5, 80–100 (1935); reprinted and revised in A. Tarski, *Logic, semantics, metamathematics* (Oxford University Press, New York, 1956), chap. 10, "Some methodological investigations on the definability of concepts."

[14] David Hilbert, *Grundlagen der Geometrie* (Stuttgart, 1899), trans. by E. J. Townsend, *The foundations of geometry* (Open Court, La Salle, Ill., 1950).

[15] This theorem is a simplified modification of Theorem 2 as presented in Tarski, "Some methodological investigations on the definability of concepts," reference 13, p. 303.

Using Padoa's method[16] we are now in a position to examine whether a given term in a deductive theory is primitive or definable. In applying Tarski's procedure on Simon's axiomatization of mechanics, let ϕ stand for the conjunction of equations (a) and (b) in Definition 3 (see p. 116), B for the set of paths of the particles for $t_0 \leqslant t \leqslant t_1$, and x for the scalars (m_1, m_2, . . . , m_n) that satisfy ϕ in terms of B. Since in general (as for instance in the case of a disjunct motion of just one particle) the numerical values of the masses are not uniquely determined, the above-mentioned theorem shows that the concept of mass in Simon's procedure is not definable in the sense of Tarski and Padoa. Thus the concept of mass, in spite of the fact that it denotes a nonobservable, should be a primitive notion.

In order to escape this unsatisfactory conclusion, Simon suggested an interesting solution of this difficulty in a paper submitted to the International Symposium on the Axiomatic Method.[17] Modifying Tarski's definition of definability, Simon introduced a definition for what he called "generic definability":

Definition: The formula $\phi(x; b_1, b_2, \ldots)$ defines generically the extralogical constant a if, for every x, if x is identical with a, x satisfies ϕ. Or in the language of symbolic logic:

$$(x) : x = a \rightarrow \phi\ (x; b_1, b_2, \ldots)$$

The introduction of the implication (in lieu of the equivalence in Tarski's expression) forecloses the demonstrability of the above-mentioned theorem and thus invalidates the application

[16] A. Padoa, "Essai d'une théorie algébrique des nombres entiers, précédé d'une introduction logique à une théorie déductive quelconque," *Bibliothèque du Congrès International de Philosophie* (Paris, 1901), vol. 3; "Un nouveau système irréductible de postulats pour l'algèbre," *Comptes rendus du Deuxième Congrès International des Mathématiciens* (Paris, 1902), pp. 249–256.

On Padoa's method see E. W. Beth, *The foundations of mathematics* (North Holland, Amsterdam, 1959), chap. 4, sec. 34; chap. 7, sec. 55; chap. 11, sec. 94. See also J. C. C. McKinsey, "On the independence of undefined ideas," *Bulletin of the American Mathematical Society 41,* 291–297 (1935); P. Suppes, *Introduction to logic* (Van Nostrand, Princeton, New York, Toronto, London, 1958), p. 169.

[17] See reference 6, p. 446.

of Padoa's method for the disqualification of the logical (generic) definability of the concept of mass.

The introduction of the concept of mass by statistical inference deserves some further remarks. No attempts to formalize Newtonian mechanics by a precise explicit definition of mass have been very successful, as we have seen. For these definitions either had to be based on the concept of force as a primitive notion or had to assume a certain dynamical law which explicitly or implicitly involved again the notion of force, in addition to the difficulty connected with the vagueness of the determinability of an appropriate inertial system. Whitehead's remark seems to be justified: "We obtain our knowledge of forces by having some theory about masses, and our knowledge about masses by having some theory about forces."[18] Although Newtonian mechanics is the simplest theory that physics has ever constructed, and although for ordinary, medium-sized physical objects Newtonian mechanics is of the highest degree of verification, its logical structure seems to defy all attempts at a complete analysis, if it is assumed that such an analysis presupposes explicit definitions of the fundamental terms involved.

In view of this situation it seems perhaps justifiable not to insist on explicit definitions of the basic terms *prior* to the construction of the theory but rather to accept the meaningfulness of these terms *through* the construction of the theory itself. In contrast to a purely hypothetico-deductive theory, as for instance axiomatized geometry, where primitive notions (like "point," "straight line," and so forth) can be taken as implicitly defined by the set of the axioms of the theory,[19] in mechanics semantic rules or correlations with experience have to be considered and a definiendum, even if defined by an implicit definition, must ultimately be determinable in its quantitative aspects through recourse to operational measurements. In fact, Mach's definition

[18] A. N. Whitehead, *An enquiry concerning the principles of natural knowledge* (Cambridge University Press, New York, 1919), p. 18.

[19] On this issue see Paul Bernays's review on Max Steck, "Ein unbekannter Brief von Gottlob Frege," *Journal of Symbolic Logic* 7, 92 (1942).

of mass already complied with this principle. Mach did not say what "mass" really is but rather advanced an implicit definition of the concept relegating the quantitative determination to certain operational procedures. Furthermore, his definition, being equivalent to Newton's third law, is part and parcel of the theory of mechanics itself and not prefatory to it. The method of statistical inference as illustrated in Simon's definition of mass is a clear and lucid expression of the idea that only through interrelations with other terms of the theory — in this case with empirically determinable accelerations — does the definiendum "mass" become meaningful.

CHAPTER 10

THE GRAVITATIONAL CONCEPT OF MASS

Definitions of mass in terms of weight are, strictly speaking, gravitational conceptions of mass. In the present chapter, however, only those concepts will be considered which characterize mass and the dynamical behavior of matter by explicit reference to a law of gravitation.

Ever since Newton in Book Three (Propositions I–VIII) of his *Principia* enunciated the law of gravitation, according to which every particle in the universe attracts every other particle with a force whose magnitude is inversely proportional to the square of their distance from each other, and "proportional to the several quantities of matter which they contain,"[1] gravitation has been regarded as a universal property of matter (just like inertia).

That the gravitational force is proportional to the mass of the attracted body (which subsequently will be called the "passive gravitational mass") Newton deduced from the experimental fact that Jupiter imparts to its satellites, the sun to the planets, and the earth to the moon and to bodies on its surface accelerations which are equal at equal distances from the respective central body. From the principle of action and reaction it then follows that the force is also proportional to the mass of the

[1] Isaac Newton, *The mathematical principles of natural philosophy,* ed. Florian Cajori (University of California Press, Berkeley, 1947), p. 414, Book 3, Proposition 7.

central body (to be called the "active gravitational mass"). Against the first part of Newton's reasoning an interesting objection was raised by E. Vicair,[2] who described Newton's contention as insufficiently established. For Newton, he says, considers only forces that act between a very large body on the one hand and a very small body on the other. But under these conditions, Vicair argues, a homogeneous function of the masses which is of a much more general structure than that proposed by Newton would equally well lead to the same results, at least to the first order of approximation. Vicair illustrates his contention by means of a Taylor expansion of such a function.

Since gravitational attraction is a universal property of matter, the law of gravitation can be used for a definition and determination of what is generally called "active gravitational mass" and will be discussed in more detail later on (pp. 126 ff).

One of the obvious methods of determining these masses of a gravitating system is that described for instance by V. V. Narlikar.[3] Let the n particles of the system whose masses $m_1, m_2, \ldots, m_n$ have to be determined be located at the time t at the points $[x_1(t), y_1(t), z_1(t)]$, $[x_2(t), y_2(t), z_2(t)]$, $\ldots$, $[x_n(t), y_n(t), z_n(t)]$ and let $a_{xk}(t)$ be the component of the acceleration of particle k along the x-axis at this instant. Then

$$a_{xk}(t) = \sum_{\substack{i=1 \\ i \neq k}}^{n} \frac{m_i[x_i(t) - x_k(t)]}{r_{ik}^3(t)}, \tag{1}$$

where $r_{ik}(t)$ is the observable distance between particle i and particle k at the time t,

$$r_{ik}(t) = \{[x_i(t) - x_k(t)]^2 + [y_i(t) - y_k(t)]^2 + [z_i(t) - z_k(t)]^2\}^{1/2}, \tag{2}$$

and the constant of gravitation is taken equal to unity. If $n - 1$ observations at $n - 1$ different instants of time furnish $n - 1$ independent Eqs. (1), the $n - 1$ mass ratios can be calculated

[2] E. Vicair, "Sur la loi de l'attraction astronomique et sur les masses des divers corps du système solaire," *Comptes rendus* 74, 790–794 (1874).

[3] V. V. Narlikar, "The concept and determination of mass in Newtonian mechanics," *Philosophical Magazine* 27, 33–36 (1939).

from the measurements of the accelerations and distances involved.

A more abstract and less elementary definition of mass in terms of gravitation was suggested in 1904 at the International Congress of Philosophy at Geneva by René de Saussure.[4] Let the vector F denote the gravitational field intensity and dS the element of a closed surface in the interior of which particles with the total mass M are enclosed. Then by Gauss's theorem,

$$\iint \mathbf{F}\, d\mathbf{S} = -4\pi G M, \tag{3}$$

where G is the constant of gravitation. If the latter is again taken as unity, Gauss's theorem can be expressed by the statement that the mass enclosed in a given surface is equal to the gravitational flux through the surface divided by 4π.

Adopting the principle of the priority of force over mass, Saussure now constructs a complete and consistent theory of mechanics on the basis of two fundamental notions: motion and constraint (*mouvement et constrainte*). If A is a geometric figure in space (point, line, and the like), a continuous extension of A in one dimension is a (one-dimensional) "sequence" of A, in two dimensions a "congruence" of A. Motion is the result of associating a sequence with a one-dimensional variable (time), constraint the result of associating a congruence with a two-dimensional variable, the flux of force (or "statoflux" in Saussure's terminology). The measure of motion is sequence divided by time (that is, velocity), the measure of constraint is flux divided by congruence (that is, force). Force is a vector normal to the congruence, velocity is tangential to sequence. Finally, mass, a derived notion, is defined by the flux (Φ) divided by 4π. Saussure then shows[5] how Newton's law of gravitation can be deduced from these assumptions and definitions.

[4] René de Saussure, "Les fondements de la mécanique," ***Rapports et comptes rendus, Congrès International de Philosophie,*** 11th session (Geneva, 1904).

[5] René de Saussure, "Les bases physiques et logiques de la mécanique," *Revue scientifique de Paris* (1905).

Newtonian mechanics, strictly speaking, distinguishes three different kinds of mass:[6] (1) *inertial mass,* which by Newton's second law of motion is determinable through its reaction to a mass-independent force; (2) *active gravitational mass,* defined as the material source of the gravitational field or the mass that "induces" gravitation, like Saussure's mass or the mass that appears in Poisson's equation; and finally (3) *passive gravitational mass,* defined as the material object of gravitational attraction or the mass susceptible to and receptive of gravitation. Traditional classical mechanics now professes a universal proportionality for all three categories of mass. The inertial mass of a body is proportional to its passive gravitational mass on the basis of empirical facts, enunciated for the first time by Newton as the result of his experiments with pendulums[7] and experimentally confirmed with increasing precision by Bessel,[8] Eötvös,[9] Pekár,[10] Southerns,[11] Zeeman,[12] and others. More recently an astronomical method, based on Kepler's so-called harmonic law, has been suggested by W. Westphal for the examination of the universal proportionality between inertial and passive gravitational mass.[13]

The universal proportionality between the active and passive gravitational masses of the same body, on the other hand, is a

[6] Cf. H. Bondi, "Negative mass in general relativity," *Reviews of Modern Physics 29,* 423–428 (1947).

[7] *Principia* (1687), p. 1: "Ponderi proportionalem esse reperi per experimenta pendulorum accuratissime instituta."

[8] F. W. Bessel, *Astronomische Nachrichten 10,* 97 (1833).

[9] R. v. Eötvös, "Untersuchungen über Gravitation und Erdmagnetismus," *Annalen der Physik 59,* 354–400 (1896); *Mathematische und naturwissenschaftliche Berichte aus Ungarn 8,* 65–103 (1890).

[10] R. v. Eötvös, D. Pekár, and E. Fekete, "Beiträge zum Gesetze der Proportionalität von Trägheit und Gravität," *Annalen der Physik 68,* 11–66 (1922).

[11] L. Southerns, "A determination of the ratio of mass to weight for a radioactive substance," *Proceedings of the Royal Society 84,* 325–344 (1910).

[12] P. Zeeman, *Proceedings of the Koninklijke Akademie van Wetenschappen te Amsterdam 20,* 542 (1917).

[13] W. Westphal, "Die Möglichkeit einer Prüfung des Satzes von der Gleichheit der trägen und der schweren Masse auf astronomischer Grundlage," *Die Naturwissenschaften 10,* 260 (1922).

consequence of Newton's third law (action = reaction), as can be seen from the following considerations. If m_{a1} and m_{p1} denote the active and passive gravitational masses of body 1 and m_{a2} and m_{p2} those of body 2 respectively, the gravitational force exerted on body 2 is given by the expression

$$F_2 = G \frac{m_{a1} m_{p2}}{r^2}$$

and the gravitational force exerted on body 1 by the expression

$$F_1 = G \frac{m_{a2} m_{p1}}{r^2},$$

where, as usual, G is the gravitational constant and r the distance between the two bodies (signs or directions are ignored). Now, according to Newton's third law $F_1 = F_2$, which implies $m_{a1}/m_{p1} = m_{a2}/m_{p2}$.

Thus, while the proportionality between inertial and passive gravitational masses is a purely empirical and accidental feature in classical physics, the proportionality between active and passive gravitational masses is deeply rooted in the very principles of Newtonian mechanics.

As H. Bondi[14] pointed out, it is a purely empirical fact that both inertial and gravitational masses are positive, or, more accurately, that they possess only one kind of polarity. It is instructive to consider the different possibilities, if this empirical fact is disregarded. If we assume, for instance, that inertial mass, as usual, has positive polarity but that gravitational masses, like electric charges, have positive or negative polarity, then obviously the response of matter to nongravitational forces would remain the familiar one while on the other hand like masses would attract and unlike masses repel each other (since Newton's law of gravitational attraction is a "negative" Coulomb law). If we take inertial mass as negative and gravitational mass as positive, matter of such a constitution would behave under the action of

[14] See reference 6.

all forces, gravitational or nongravitational, in exactly the reverse of the way it does in nature known to us.

The existence of negative gravitational masses has been seriously considered as a solution of certain difficulties when Newton's law of gravitation was applied to the universe as a whole. The first to call attention to such difficulties was probably Carl Neumann.[15] The same problem was twenty years later restated in a mathematically more rigorous way by H. Seeliger.[16] The number of lines of force converging from infinity on a body of (passive gravitational) mass m is proportional to m (in analogy to electrostatics). Let us assume that the matter of the universe is of uniform distribution with average density ρ and unbounded in spatial extension; then $\frac{4}{3}\pi\rho r^3$ lines cross the surface of a sphere of radius r. Since the area of the surface is $4\pi r^2$, the density of lines of force, that is, the gravitational intensity, is $\frac{1}{3}\rho r$ and thus increases in direct proportion to the radius of the sphere. As the center of the latter can be chosen arbitrarily far away, the intensity of the gravitational field would necessarily be infinite at any given point in space. In view of this paradoxical situation Seeliger concluded: "Newton's law, undoubtedly, is not rigorously accurate but has to be modified by certain terms whereby those difficulties are removed."[17] Seeliger's first suggestion of such "supplementary terms" (*Ergänzungsglieder*) was the introduction of an exponential (absorptive) factor, in analogy to Laplace's "loi de diminution";[18] Newton's law, according to Seeliger, should thus read:

[15] Carl Neumann, "Über die den Kräften elektrodynamischen Ursprungs zuzuschreibenden Elementargesetze," *Abhandlungen der math.-phys. Classe der köngl. sächsischen Gesellschaft der Wissenschaften 10,* 417–524 (Leipzig, 1874).

[16] H. Seeliger, "Über das Newton'sche Gravitationsgesetz," *Astronomische Nachrichten 137,* 129–136 (1895); "Über das Newton'sche Gravitationsgesetz," *Sitzungsberichte der math.-phys. Classe der köngl. bayrischen Akademie der Wissenschaften zu München 26,* 373–400 (1896); see also *Astronomische Nachrichten 138* (No. 3292).

[17] *Astronomische Nachrichten 137,* 134 (1895).

[18] Pierre Simon Marquis de Laplace, *Traité de mécanique céleste,* book 16, chap. 4, sec. 6; in *Oeuvres complètes* (Académie des Sciences, Paris, 1882), vol. 5, p. 448.

$$F = G\frac{m_1 m_2}{r^2}e^{-\lambda r}.$$

Seeliger's suggestion was soon rejected for the following reason. The numerical value of the correction coefficient could be computed from the precession of the perihelion of Mercury; this value, however, if used for the computation of the precession of the perihelion of other planets, led to results inconsistent with observational data. Numerous other modifications of Newton's formula have subsequently been proposed, yet all met with little success. In view of these failures, August Föppl made a different attempt to overcome these difficulties. In an article entitled "On a possible generalization of Newton's law of gravitation[19] Föppl introduced the notion of "negative mass." In fact, Karl Pearson, in his "hydronomical" explanation of gravitation and magnetism, had already spoken of attractive and repulsive gravitational forces[20] and of "inverted magnetism." But it was Föppl who developed a logically consistent theory of positive and negative masses in analogy to Maxwell's theory of the electromagnetic field of positive and negative charges.

In analogy to the electric field vector **E**, the gravitational field at a given point in space is determined by an irrotational vector $\mathbf{V} = -\text{grad}\ \varphi$, where φ is the gravitational potential at that point and equal to $-Gm/r$. Maxwell's well-known expression for the energy density of the field, $\frac{1}{2}\epsilon_0\mathbf{E}^2$, in view of the "negative" Coulomb law of gravitation, has to be modified to $e_0 - \frac{1}{2}c\mathbf{V}^2$, where the constant e_0 is the energy density for $\mathbf{V} = 0$ and $c = (1/4\pi)G$. The necessity of introducing e_0 for the case of a "nega-

[19] August Föppl, "Über eine mögliche Erweiterung des Newton'schen Gravitations-Gesetzes," *Sitzungsberichte der math.-phys. Classe der K. B. Akademie der Wissenschaften zu München 27,* 93–99 (1897).

[20] Karl Pearson, "On the motion of spherical and ellipsoidal bodies in fluid media," *Quarterly Journal of Pure and Applied Mathematics 20,* 60–80, 184–211 (1885). See also K. Pearson, "Ether squirts," *American Journal of Mathematics 13,* 309–362 (1891), where he speaks of matter found in equal and opposite quantities of two kinds, and where he thinks it possible that the fast-receding "1830 Groombridge" star, the star with the greatest proper motion known at that time, is composed of negative mass and thus repelled from our region in space.

tive" Coulomb law was pointed out already by Maxwell.[21] The sum total of gravitational energy in the universe is then given by the expression

$$T = T_0 - \tfrac{1}{2}c \iiint \mathbf{V}^2 \, d\tau,$$

where

$$T_0 = \iiint e_0 \, d\tau$$

and the integration extends over the whole of space. By Green's theorem,

$$\iiint \mathbf{V}^2 \, d\tau = 4\pi \iiint \varphi\rho \, d\tau,$$

where ρ is the source density and $d\tau$ (that is, the active gravitational mass contained in $d\tau$).[22]

Let us now assume, says Föppl, that ρ, or for point masses q, can be either positive or negative. All our conclusions are still valid. A displacement of the point mass q_1 by $\delta\mathbf{r}$ changes the total energy T of the field by

$$\delta T = cq_1 \iiint \frac{\rho}{r^3} \mathbf{r} \, d\tau \, \delta\mathbf{r}.$$

If F is the force necessary for the displacement, the conservation of energy requires that

$$\mathbf{F} \, \delta\mathbf{r} + \delta T = 0$$

or

$$\mathbf{F} = -c \iiint \frac{q_1\rho}{r^3} \mathbf{r} \, d\tau,$$

[21] James Clerk Maxwell, *Scientific papers* (Cambridge, 1890), vol. 1, p. 570. The article referred to was originally published in *Transactions of the Royal Society* (*London*) *155*, 492 (1865).

[22] Since $4\pi\rho = \text{div } \mathbf{V}$ and $\mathbf{V} = -\text{grad } \varphi$,

$$\iiint \varphi\rho \, d\tau = \frac{1}{4\pi} \iiint \varphi \, \text{div } \mathbf{V} \, d\tau = \frac{1}{4\pi} \iiint [\text{div } (\varphi\mathbf{V}) - \mathbf{V} \, \text{grad } \varphi] \, d\tau$$
$$= \frac{1}{4\pi} \left\{ \iint \varphi V_n \, d\sigma + \iiint \mathbf{V}^2 \, d\tau \right\} = \frac{1}{4\pi} \iiint \mathbf{V}^2 \, d\tau.$$

or for two point masses

$$\mathbf{F} = -c\,\frac{q_1 q_2}{r^3}\,\mathbf{r},$$

which is Newton's law of gravitation. By the introduction of negative masses as sinks of gravitational lines of force, in contrast to positive masses as their sources, the above-mentioned difficulty is, of course, removed, for each line of force extends only from source to sink. The empirical absence of mutually repelling masses, Föppl argues, can be explained by the plausible assumption that the negative masses, repelled by the positive masses prevalent in our region of space, have been driven away to distances inaccessible to our experience.

Föppl's hypothesis leads to an interesting result: the constant e_0, in virtue of the requirement that masses of the same kind attract each other, is necessarily at least as great as the maximum value of $|\frac{1}{2}c\mathbf{V}^2|$, and thus in interstellar space, where approximately $\mathbf{V} = 0$, the energy density becomes "exorbitantly" high.

Föppl was fully aware that his notion of mass is fundamentally different from the conception of inertial mass.

> Whenever I speak of masses [he remarks], I mean by this term not the quantity that appears in the law of inertia or in the fundamental equation of motion, but rather the material substratum of the universe as far as it constitutes the carrier of gravitational phenomena. "Mass," consequently, is in this context as in the theory of electricity synonymous with source of flux of force.[23]

The notion of negative gravitational mass, as introduced by Pearson and Föppl, soon became the subject of profuse philosophical speculations. Schuster, for instance, in a paper entitled "Potential matter — a holiday dream," published in *Nature,*[24] imagined a world in which atoms are sources through which an invisible fluid enters the three-dimensional space. The constant creation of this fluid in these sources is compensated by its disap-

[23] Reference 19, p. 96.
[24] *Nature 58,* 367 (1898).

pearance in an equal number of sinks. "These sinks would form another set of atoms, possibly equal to our own in all respects but one: they would mutually gravitate towards each other, but be repelled from the matter which we deal with on this earth." Schuster then speaks of these "antiatoms" as "antimatter" and envisages the possibility that through the fusion of matter with antimatter not only gravity but also inertia may be neutralized and some kind of "potential matter" produced in analogy to the disappearance of kinetic energy and its storage as potential energy. "Is our much-exalted axiom of the constancy of mass an illusion based on the limited experience of our immediate surroundings?" he concludes his speculations.

For the modern physicist who studies the cosmological hypotheses of continuous creation and the nuclear dynamics of antiparticles, conjectures such as Schuster's may seem to have been more than reveries, although, of course, Schuster's concept of antimatter is not identical with the modern concept of antiparticle. And yet, one should not overestimate the scientific importance of such speculations. However, the results of serious scientific research often do not fall short of the most fantastic flight of fancy. Thus F. A. Kaempffer in a recent investigation concerning Mach's principle,[25] following certain suggestions by P. Morrison and T. Gold, considers the possibility that in addition to the positive gravitational rest mass of matter and the negative gravitational rest mass of antimatter there exists a third kind of gravitational mass, not the "potential mass" of Schuster, but the "kinetic mass" which has the same sign both for matter and for antimatter and which vanishes both for matter and for antimatter "at rest." More precisely, it is the interaction of the kinetic masses of the universe by which the inertia of both matter and antimatter is supposed to be accounted for in conformity with Mach's cosmological principle.

[25] F. A. Kaempffer, "On possible realizations of Mach's program," *Canadian Journal of Physics 36,* 151–159 (1958).

It should, however, be noted that L. I. Schiff in a recent article on "Gravitational properties of antimatter"[26] advanced a number of important arguments in favor of the assumption that *all* particles, of matter as well as of antimatter, have positive rest masses and positive passive gravitational masses.

If, according to Mach's principle, inertia is an effect depending upon the distribution of matter at large, any asymmetry in the allocation of matter in space should give rise to an isotropy of inertia: the inertial behavior of macroscopic and microscopic bodies should exhibit directional differentiations. Our solar system, as we know, occupies a pronounced peripheral position with respect to the Galaxy. It is therefore most probable — always provided that Machs' principle holds — that the inertial response of terrestrial objects or test particles is a function of their direction with respect to the center of the Galaxy. Mathematically speaking, mass ceases to be a scalar and becomes a tensor quantity in its relation to the directional vector quantities of force and acceleration.

The advantage of regarding mass as a tensor was suggested — though in a different context — in 1945 by van den Dungen,[27] who recommended the representation of mass as a tensor of the second rank primarily in the interest of notational consistency. Van den Dungen pointed out that the fundamental equation of classical mechanics, $md^2x^i/dt^2 = F^i$, is usually written in terms of contravariant vectors, and correctly so since the displacements dx^i, by definition, are contravariant vectors. On the other hand, the scalar quantity work is defined by the equation $dW = F_i\, dx^i$ in terms of the covariant components of the force vector. Thus a metric is required to transform contravariant into covariant components.

If mass is defined as a tensor m_{ij}, the covariant momentum p_i

[26] *Proceedings of the National Academy of Sciences, Washington, D.C. 45*, 69–80 (1959).

[27] F. H. van den Dungen, "Sur la notion de masse," *Académie Royale de Belgique, Bulletin de la classe des sciences 31*, 666–668 (1945).

can be expressed by the equation $p_i = m_{ij}\, dx^j/dt$ (summation over the repeated index j) and the fundamental law of motion can be written as $dp_i/dt = F_i$. If any directional effect of the Machian action is completely disregarded, as in van den Dungen's paper, the mass tensor reduces for rectangular coordinate systems to a diagonal tensor $m_{ij} = m\delta_{ij}$ (δ_{ij} are the Kronecker symbols). The kinetic energy of a particle is expressed as the invariant $\frac{1}{2}m_{ik}(dx^i/dt)(dx^k/dt)$ and its transformation to oblique coordinates is simplified. The procedure can easily be generalized for the relativistic velocity-dependent mass. But apart from such an increase in notational consistency the procedure advocated gives no further insight into the physical significance of the concept of mass.

If, however, mass is represented as a tensor in order to investigate experimentally a possible directional effect in the inertial response of matter, a positive result of such an experiment would not only demonstrate the validity of the controversial Machian principle but would also throw new light on the concept of mass as an inductive effect.

In search for an experiment to detect the anisotropy of inertia G. Cocconi and E. Salpeter[28] assume that the contribution of galactic matter to the inertial mass of terrestrial objects is a maximum if these objects are accelerated in the direction to (or from)[29] the Galactic Center and is a minimum if they are accelerated perpendicularly to this direction. The mass tensor m_{ij} is characterized by the direction of the three principal axes and the corresponding diagonal elements m_{ii}. If the diagonal element corresponding to the principal axis in the direction to the galactic center is taken as $m + \Delta m$ (and the diagonal elements for the other two axes as $m - \frac{1}{2}\Delta m$, the trace being m), the anisotropic

[28] G. Cocconi and E. Salpeter, "A search for anisotropy of inertia," *Nuovo cimento 10*, 646–651 (1958).

[29] If the inertial mass for forward motion toward the Galactic Center differed from the inertial mass for motion in the opposite direction, ordinary forces would cease to be conservative.

contribution $\Delta m/m$ can be estimated on the basis of a quantitative formulation of Mach's principle. For this purpose Cocconi and Salpeter assume that this contribution depends linearly on the inertia-producing mass M and is inversely proportional to the kth power of the distance involved. Furthermore, a limited range of action must be posited (corresponding to the "radius R of the universe" in cosmology). If ρ is the average density of matter within this sphere of action, then obviously the isotropic part m of the inertia is proportional to

$$\int_0^R \frac{4\pi r^2 \rho \, dr}{r^k} = \frac{4\pi\rho}{3-k} R^{3-k}.$$

If in addition a mass M, at a distance r from the test particle, causes an anisotropic contribution Δm, then clearly

$$\frac{\Delta m}{m} = \frac{M}{r^k} \frac{3-k}{4\pi\rho R^{3-k}}.$$

With $\rho = 10^{-28}$ gm cm^{-3}, $R = 10^{28}$ cm ($= cT$ where T^{-1} is Hubble's constant and c the velocity of light) and assuming $k = 1$, the value of $\Delta m/m$ is of the order of 10^{-7} with respect to the Galaxy; if $k = 0.25$, $\Delta m/m \sim 10^{-11}$.

In principle, such an effect should be observable in numerous macroscopic phenomena, as for instance a diurnal variation of the period of a quartz-crystal clock. Since, however, present-day precision is still below the level necessary for the detection of such effects, as far as macroscopic observations are concerned, Cocconi and Salpeter suggest a microwave measurement of the Zeeman splitting of single electron levels with different orientations of the magnetic field relative to the Galactic Center. In these measurements a precision of the order of 1 part in 10^{11} should be attainable. A. Carelli[30] suggests optical precision measurements of double refraction phenomena in liquids or gases

[30] A. Carelli, "On Mach's principle," *Nuovo cimento 13,* 853–856 (1959).

in the visible region (involving two perpendicular planes of motion of the electrons). Unfortunately, as yet no decisive experiments[31] seem to have been carried out and the problem of the validity of Mach's principle as well as the interpretation of mass according to this principle still remain intriguing questions.

[31] G. Cocconi and E. Salpeter recently suggested the use of the Mössbauer effect in order to increase the sensitivity of measuring $\Delta m/m$ to 10^{-14}. See *Physical Review Letters 4,* 176–177 (1960).

CHAPTER 11

THE ELECTROMAGNETIC CONCEPT OF MASS

The idea that inertia is ultimately an electromagnetic phenomenon and that inertial mass is basically an inductive effect had its origin in the study of the electrodynamics of charges in motion. Although Maxwell's electromagnetic stress tensor, as the spatial part of the energy-momentum tensor of the electromagnetic field, contained implicit ideas conducive to this new conception — as is known today but was unknown before the rise of relativity — it was only in 1881 that Joseph John Thomson, in an article "On the electric and magnetic effects produced by the motion of electrified bodies"[1] envisaged the possibility of reducing inertia to electromagnetism. Thomson considered the case of an electrostatically charged sphere moving through an unlimited medium of specific inductive capacity (dielectric constant) ϵ. In order to calculate the displacement current produced by the motion of the charge, Thomson assumed (1) that the surface charge distribution on the conductor remains unaltered by the motion and (2) that the electric field is carried forward with the moving conductor undistorted. That the first assumption was legitimate and the second fallacious was shown only fifteen years later by W. B. Morton.[2]

Since according to Maxwell's theory a variation of the electric

[1] *Philosophical Magazine 11,* 229–249 (1881).
[2] *Philosophical Magazine 41,* 488 (1896).

displacement (that is, a displacement current) causes the same effects as an ordinary electric current, a magnetic field is produced which can be computed from the vector potential corresponding to the displacement current. From the electric and magnetic field intensities Thomson thus was able to calculate the energy of the surrounding electromagnetic field. This energy, according to the conservation principle, must have been supplied by the motion of the charged conductor. Since its motion serves as a source of energy it is clear that the conductor must experience a resistance as it moves through the dielectric. Frictional dissipation of energy through the medium being excluded, as the medium is assumed to have no conductivity, the resistance experienced must be analogous to that of a solid moving through a perfect fluid. "In other words, it must be equivalent to an increase in the mass of the charged moving sphere." [3]

For Thomson, as we see, the resulting resistance experienced is only "equivalent to an increase in mass"; he conceives the process only "as if" the mass increased. He still thinks in analogy to classical hydrodynamics; here a spherical particle of mass m, immersed in an incompressible fluid through which it moves with velocity v, acquires, in addition to its own kinetic energy $\frac{1}{2}mv^2$, also the energy $\frac{1}{2}\mu v^2$; the total work communicated to the whole system can thus be written as $\frac{1}{2}(m + \mu)v^2$. "The presence of the liquid has therefore the apparent effect of increasing the mass of the sphere," [4] and μ is in fluid mechanics often referred to as the "induced mass." [5]

As Thomson showed in his article, the virtual increase of mass μ in the case of a sphere of radius a and charge e is given by the expression

$$\mu = \frac{4}{15}\frac{e^2}{ac^2}. \qquad (1)$$

[3] Reference 1, p. 230.

[4] D. E. Rutherford, *Fluid mechanics* (Oliver and Boyd, Edinburgh and London; Interscience Publishers Inc., New York, 1959), p. 103.

[5] See, for instance, R. J. Seeger, "Fluid mechanics," in *Handbook of physics*, ed. E. U. Condon and H. Odishaw (McGraw-Hill, New York, Toronto, London, 1958), pp. 3–18.

For an evaluation of the order of magnitude of μ with respect to the ordinary inertial mass m of the body in motion, Thomson computed the apparent increase in mass of the earth in its orbit around the sun on the assumption that the earth is charged to the highest possible potential. He showed that μ amounts to only 7×10^8 gm, "a mass, which is quite insignificant when compared with the mass of the earth." [6] We quote this example and this quotation in order to show that Thomson, at this stage, is still far from generalizing his conclusions and interpreting all inertial mass as "induced mass." It is also important to note that according to Thomson's calculations μ does not depend on the velocity v of the moving body.

Immediately after the publication of Thomson's paper George Francis FitzGerald pointed out[7] that Thomson's displacement current as used in his calculations does not comply with Maxwell's circuital condition and that only the current composed of this displacement current and the convection current originated by the moving charge itself does satisfy this condition.

An important improvement over Thomson's result was Oliver Heaviside's investigation "On the electromagnetic effects due to the motion of electrification through a dielectric," [8] published in 1889, which led to a result that included terms of higher order in v/c. In a modernized version Heaviside's solution may be presented as follows. A point charge q moves in the positive direction of the x-axis of a coordinate system (x, y, z) with constant velocity v. In order to calculate the energy of the electromagnetic field due to the motion alone, that is, the energy excess over the electrostatic-field energy corresponding to a stationary charge, Maxwell's theory requires the computation of $(1/8\pi) \iiint \mathbf{H}^2 \, d\tau$, where $\mathbf{H}$ is the magnetic-field vector to be determined by Max-

[6] Reference 1, p. 234.

[7] George Francis FitzGerald, *The scientific writings of the late George Francis FitzGerald, collected and edited with a historical introduction by Joseph Larmor* (Longmans, Green, London, 1902), p. 102. The original paper appeared in the *Proceedings of the Royal Dublin Society 3,* 250 (1881).

[8] *Philosophical Magazine 27,* 324–339 (1889).

well's equations from the convection and displacement currents associated with the motion of the point charge. Since the velocity vector v points in the x-direction, it is easy to show that Maxwell's equations impose on the vector potential **A** the following conditions:

$$\frac{1}{c^2}\frac{\partial^2 A_x}{\partial t^2} - \nabla^2 A_x = \frac{4\pi}{c}\rho v, \quad A_y = 0, \quad A_z = 0, \tag{2}$$

where ρ is the volume density of the charge, which in modern notation may be taken as related to q by the equation $\rho = q\delta(r - r_0)$, where r_0 is the position vector of the charge and δ denotes the Dirac function. By a transformation to a new coordinate system (ξ, η, ζ) which is attached to the moving charge and contracted in the direction of motion by the factor $(1 - v^2/c^2)^{-1/2}$, so that

$$\xi = \left(1 - \frac{v^2}{c^2}\right)^{-1/2}(x - vt), \quad \eta = y, \quad \zeta = z,$$

the preceding equation for A_x turns into a simple Poisson equation

$$\frac{\partial^2 A_x}{\partial \xi^2} + \frac{\partial^2 A_y}{\partial \eta^2} + \frac{\partial^2 A_z}{\partial \zeta^2} = -\frac{4\pi}{c}\rho v, \tag{3}$$

which can be solved for A_x. For a single charge q the x-component of the vector potential A_x, if transformed back into the original coordinate system (x, y, z), is given by the expression

$$A_x = \frac{q\beta}{[x^2 + (1 - \beta^2)(y^2 + z^2)]^{1/2}},$$

where $\beta = v/c$. Once the components of the vector potential are known, the equation

$$\mathbf{H} = \text{curl } \mathbf{A}$$

determines the value of the components of the magnetic-field vector **H**. It is then easy to verify that

$$H^2 = \frac{q^2(1 - \beta^2)\beta^2(y^2 + z^2)}{[x^2 + (1 - \beta^2)(y^2 + z^2)]^3} \tag{4}$$

or

$$H^2 = (q^2\beta^2 \sin^2 \theta)/r^4,$$

where the spherical coordinates r and θ are defined by the relations

$$r^2 = x^2 + y^2 + z^2, \quad \sin^2 \theta = \frac{y^2 + z^2}{x^2 + y^2 + z^2},$$

and only terms of the order of β^2 are retained.

Heaviside now assumed — erroneously, as G. F. C. Searle later pointed out[9] — that the foregoing computation of H^2 remains valid also when q is not interpreted as the charge of a point charge but as the charge distributed over the surface of a perfectly conducting sphere of radius a. The additional energy ΔU of the electromagnetic field outside the moving sphere, due to its motion, is thus given by

$$\Delta U = \frac{1}{8\pi} \iiint H^2 \, d\tau = \frac{q^2\beta^2}{8\pi} \int_a^\infty \int_0^{2\pi} \int_0^\pi \frac{\sin^2 \theta}{r^4} r^2 \sin \theta \, d\theta \, d\varphi \, dr$$

or finally[10]

$$\Delta U = \frac{q^2 v^2}{3ac^2}. \tag{5}$$

Comparing this expression with $\frac{1}{2}\mu v^2$, Heaviside obtained for the increase in mass the value

$$\mu = \frac{2q^2}{3ac^2}. \tag{6}$$

[9] G. F. C. Searle, "On the steady motion of an electrified ellipsoid," *Philosophical Magazine 44*, 329–341 (1897). See also Searle, "Problems in electric convection," *Philosophical Transactions of the Royal Society of London* [A] *187*, 675–713 (1897).

[10] Equation (5) can be derived in an elementary way from the law of Biot and Savart (sometimes referred to also as Laplace's law or Ampère's law), according to which the magnetic field intensity (in Gaussian units) at a distance r from the charge q, moving with velocity v, is $H = qv \sin \theta/cr$, where θ is the angle between $\mathbf{v}$ and $\mathbf{r}$. The field energy dU in the volume element $d\tau$ is given by $dU = (1/8\pi)H^2 \, d\tau$. Substituting the value of H and integrating over all space outside the spherical particle yields Eq. (5). If the charge q were assumed to be distributed throughout the volume of the sphere, the result would have to be changed by a factor of the order of magnitude of unity.

Since the total electromagnetic energy U_0 outside a stationary sphere with surface charge q and radius a is $q^2/2a$, as can be shown by a simple integration, Heaviside's result may be expressed as follows: The increase in mass of a moving sphere with uniform surface charge distribution is 4/3 of its stationary field energy U_0 divided by c^2.

For Heaviside, in contrast to Thomson, this increase of mass is a physically significant phenomenon, not only analogous to mechanical inertia but an inertial effect *sui generis*. In fact, Heaviside speaks explicitly of an "electric force of inertia."[11]

The publication of Heaviside's article marked the beginning of an animated competition between the science of mechanics and the science of electromagnetism for the primacy in physics. The era of mechanical interpretations of electromagnetic phenomena initiated by William Thomson (Lord Kelvin) and Maxwell in their search for mechanical models of the ether was still at its peak. The scientific journals of the nineties were flooded with articles that tried to reduce electromagnetism to mechanics or hydrodynamics. The belief that all forces in nature are ultimately only different manifestations of the same fundamental power inspired many scientists to search for such principles of unification. Thus, for instance, A. Korn's mechanical theory of the electromagnetic field, based on C. A. Bjerknes's theory of pulsating spheres,[12] and originally published as a book under the title *A theory of gravitation and electrical phenomena on the basis of hydrodynamics*,[13] gained considerable popularity in the world of science at that time. As late as 1917 and 1918 the *Physikalische Zeitschrift* was still publishing almost unremittingly papers by Korn and others on this and similar theories.[14]

[11] Reference 8, p. 332.

[12] *Göttinger Nachrichten* (1876), p. 245; *Comptes rendus 84*, 1375 (1877); *Nature 24*, 360 (1881).

[13] A. Korn, *Eine Theorie der Gravitation und der elektrischen Erscheinungen auf Grundlage der Hydrodynamik* (Berlin, ed. 1, 1894; ed. 2, 1896; ed. 3, 1898).

[14] We mention here only Korn's papers during 1917 and 1918: "Mechanische Theorien des Elektromagnetischen Feldes," *Physikalische Zeitschrift 18*, 323–326, 341–345, 504–507, 539–542, 581–584 (1917); *19*, 10–13, 201–203, 234–237, 426–429 (1918).

These mechanical theories of electromagnetism had now to face the rivalry of electromagnetic theories of mechanics. The search for conceptual unification motivated many theorists at the turn of the century to explore the methodological possibilities of the new approach based on the electromagnetic conception of mass. Furthermore, thoughtful theorists may perhaps have regarded the enormous abundance of publications propounding mechanical theories of electromagnetism as symptomatic of the ultimate futility of all attempts in this direction and welcomed the new approach in the opposite direction. Thus, for instance, Ludwig Boltzmann, in his *Lectures on the principles of mechanics,* with reference to the newly developed theory of electrons, remarked:

> This theory certainly does not intend to explain the concepts of mass and force, the law of inertia, and so forth, from something more simple and something that can be comprehended more easily; its fundamental conceptions and basic laws will undoubtedly remain as inexplicable as those of the mechanical point of view. However, the advantage of deriving the whole science of mechanics from conceptions which anyhow are indispensable for the explanation of electromagnetism would be as important as if conversely electromagnetic phenomena were explained on the basis of mechanics. May the former succeed."[15]

One of the early enthusiastic proponents of the electromagnetic concept of mass was Wilhelm Wien. In an article "On the possibility of an electromagnetic foundation of mechanics,"[16] published in the famous Lorentz jubilee volume, Wien first admits that Maxwell, Kelvin, and Hertz have chosen the natural way and taken mechanics as the basis for the derivation of Maxwell's equations. However, in view of the ever-increasing complexity of the mechanical models proposed for this purpose, Wien now thinks it more promising for the future development of physical

[15] Ludwig Boltzmann, *Vorlesungen über die Prinzipe der Mechanik* (Leipzig, 1897); quoted from the second edition (Leipzig, 1904), part 2, pp. 138–139.

[16] Wilhelm Wien, "Über die Möglichkeit einer elektromagnetischen Begründung der Mechanik," in *Recueil de travaux offerts par les auteurs à H. A. Lorentz* (The Hague, 1900), pp. 96–107.

theory to consider the electromagnetic equations as the basis for the derivation of the laws of mechanics.[17]

Referring to Heaviside's "electric force of inertia," Wien generalizes this result and expresses his firm conviction in the derivability of mechanical inertia from electromagnetic theory. "The inertia of matter, which apart from gravitation gives an independent definition of mass, can be deduced without additional hypotheses from the already frequently employed notion of electromagnetic inertia." [18]

Wien's derivation of inertial mass is based on Searle's computation concerning the field energy produced by a charged "Heaviside ellipsoid" in motion.[19] If U is the energy of the field corresponding to such an ellipsoid moving with velocity v, and if U_0 is the energy corresponding to the ellipsoid at rest, Wien concludes on the basis of Searle's results that

$$U = U_0 \frac{1 + \frac{1}{3}\beta^2}{(1 - \beta^2)^{1/2} \arc \sin \beta}. \tag{7}$$

Expanding this expression in terms of β, Wien obtains

$$U = U_0 \left(1 + \frac{2}{3}\beta^2 + \frac{16}{45}\beta^4 + \cdots\right). \tag{8}$$

The increase in energy, due to the motion of the charge, is consequently to a first approximation

$$\frac{2}{3} U_0\beta^2 = \frac{2}{3c^2} U_0 v^2 = \frac{1}{2} \mu v^2 \tag{9}$$

[17] "Viel aussichtsvoller als Grundlage für weitere theoretische Arbeit scheint mir der umgekehrte Versuch zu sein, die elektromagnetischen Grundgleichungen als die allgemeineren anzusehen, aus denen die mechanischen zu folgern sind." *Ibid.*, p. 97.

[18] "Die Trägheit der Materie, welche neben der Gravitation die zweite unabhängige Definition der Masse giebt, lässt sich ohne weitere Hypothesen aus dem bereits vielfach benutzten Begriff der elektromagnetischen Trägheit folgern." *Ibid.*, p. 101.

[19] This term, introduced by Searle (see reference 9), denotes an oblate spheroid whose major axes have the ratios $(1 - \beta^2):1:1$. When moving with velocity v, its field is identical with that of a point charge having the same charge and the same velocity.

and the inertial mass

$$\mu = \frac{4}{3}\frac{U_0}{c^2}. \tag{10}$$

Wien thus confirms Heaviside's result for small velocities v but shows that for higher velocities other terms of the expansion have to be taken into account: the electromagnetic mass is velocity-dependent.

The subsequent development of the electromagnetic concept of mass, especially as expounded by its most eloquent advocate, Max Abraham, is intimately connected with Poynting's discovery in 1884 of his famous theorem[20] concerning the transfer of energy in the electromagnetic field and is also intimately connected with the concept of the electromagnetic momentum whose theoretical importance had already been foreshadowed by Poincaré[21] but was worked out in full detail only by Abraham himself.[22]

Abraham's study of the electromagnetic nature of inertial mass was confined to the mechanics of the electron, which at that time, in particular through the work of W. Kaufmann, was the subject of much experimental research. It was, however, tacitly assumed that the conclusions arrived at would be applicable to positive charges as well and would thus furnish a theory of matter in general. In a chapter entitled "The fundamental hypotheses of the dynamics of the electron and the electromagnetic world-picture" of his famous textbook on *The theory of electricity*[23]

[20] John Henry Poynting, *Philosophical Transactions 175,* 343 (1884). "Poynting's theorem" was independently discovered by Heaviside; see *The Electrician 14,* 178, 306 (1885). In Russian scientific literature the theorem is often associated with the name of N. A. Umov.

[21] Henri Poincaré, "La théorie de Lorentz et le principe de réaction," *Recueil de travaux offerts par les auteurs à H. A. Lorentz,* pp. 252–278, particularly pp. 276–277.

[22] The concept "electromagnetic momentum" (*elektromagnetische Bewegungsgrösse*) was introduced by Abraham in his article "Die Dynamik des Elektrons," published in the *Göttinger Nachrichten* (1902), pp. 20–41, and fully employed for the first time in his paper "Prinzipien der Dynamik des Elektrons," *Annalen der Physik 10,* 105–179 (1903).

[23] Max Abraham, *Theorie der Elektrizität* (Teubner, Leipzig, 1905), vol. 2, sec. 16, "Die Grundhypothesen der Dynamik des Elektrons und das elektromagnetische Weltbild," p. 139.

Abraham stated that the objective of his investigations is the development of a dynamics of the electron that could account for Kaufmann's experiments on a purely electromagnetic basis.

Starting from Maxwell's equations and from the so-called Lorentz formula for the force density f (the force per unit volume exerted by the field on the material charges),

$$\mathbf{f} = \rho\left(\mathbf{E} + \frac{1}{c}\mathbf{v} \times \mathbf{H}\right). \tag{11}$$

Abraham shows that the x-component, for example, of the force density is given by the expression

$$f_x = \frac{\partial T_{xx}}{\partial x} + \frac{\partial T_{xy}}{\partial y} + \frac{\partial T_{xz}}{\partial z} - \frac{dg_x^{(f)}}{dt}, \tag{12}$$

where T_{xx}, T_{xy}, . . . are the components of the electromagnetic stress tensor and $\mathbf{g}^{(f)}$ is the electromagnetic momentum density of the field, that is,[24]

$$\mathbf{g}^{(f)} = \frac{1}{4\pi c}\mathbf{E} \times \mathbf{H}. \tag{13}$$

The total force $\mathbf{F}$ exerted by the field on the material system is the volume integral over the force density,

$$\mathbf{F} = \iiint \mathbf{f}\, d\tau, \tag{14}$$

and, according to Newton's law, can be expressed as the time derivative of the total material or mechanical momentum $\mathbf{G}^{(m)}$. Consequently the x-component of $\mathbf{F}$ satisfies the equation

$$\frac{dG_x^{(m)}}{dt} = \iiint \left(\frac{\partial T_{xx}}{\partial x} + \frac{\partial T_{xy}}{\partial y} + \frac{\partial T_{xz}}{\partial z}\right) d\tau - \frac{dG_x^{(f)}}{dt}, \tag{15}$$

where $\mathbf{G}^{(f)}$, the total electromagnetic momentum of the field, is the volume integral of the field momentum density. By means of Gauss's divergence theorem it can be shown that the integral over the tensor divergence vanishes (if the boundary surfaces

[24] As this equation indicates, the Poynting vector is a carrier not only of energy but also of momentum, a fact which, as we shall see in the following two chapters, found its explanation only in the special theory of relativity.

are chosen sufficiently far removed). Abraham thus obtains the conservation law of linear momentum for the electromagnetic field:

$$\frac{d\mathbf{G}^{(m)}}{dt} = -\frac{d\mathbf{G}^{(f)}}{dt}. \tag{16}$$

Abraham next computes the field momentum $\mathbf{G}^{(f)}$ for an electron moving with velocity $\mathbf{v}$ along the positive x-axis of a stationary coordinate system. Using the law of Biot and Savart, $\mathbf{H} = (1/c)\,\mathbf{v} \times \mathbf{E}$ and elementary vector identities, he obtains for the x-component

$$g_x^{(f)} = \frac{v}{6\pi c^2} E^2, \tag{17}$$

and, since by symmetry considerations $G_x^{(f)}$ and $G_y^{(f)}$ are equal to zero, for the total field momentum,

$$\mathbf{G}^{(f)} = \frac{\mathbf{v}}{6\pi c^2} \iiint E^2\, d\tau. \tag{18}$$

Thus, if U_0 denotes the total energy of the field, namely, $(1/8\pi) \iiint E^2\, d\tau$,

$$\mathbf{G}^{(f)} = \frac{4}{3} \frac{U_0}{c^2} \mathbf{v}. \tag{19}$$

From these calculations Abraham draws the following conclusions. If the velocity of the electron $\mathbf{v}$ is constant in magnitude and in direction, $\mathbf{G}^{(f)}$ is likewise a constant and its time derivative is zero. From Eq. (16), the law of conservation of linear momentum of matter and field, it then follows that $\mathbf{G}^{(m)}$ is also a constant. This, according to Abraham, is the electromagnetic interpretation of the law of inertia. In his fundamental paper on the "Principles of the dynamics of the electron" Abraham gives an explicit formulation of the electromagnetic version of the law of inertia as follows:

If, from the beginning, the motion of the electron was uniform and purely translational, and if its velocity was less than the velocity of light, then for the continuation of uniform motion no external forces or torques are required.[25]

If the velocity **v** of the electron increases or decreases without change in direction, the vector $d\mathbf{G}^{(f)}/dt$ likewise remains in the direction of the motion while its magnitude is given by

$$\frac{d\mathbf{G}^{(f)}}{dt} = \frac{dG^{(f)}}{dv}\frac{d\mathbf{v}}{dt} = \mu\mathbf{w}, \tag{20}$$

where **w** is the kinematic acceleration and μ is the electromagnetic mass. From Eq. (16) it then follows that the electron is subject to a force which acts in the direction opposite to the motion and which is equal in magnitude to the acceleration multiplied by the electromagnetic mass.

Abraham regards the electron as a rigid sphere with a homogeneous distribution of charge (either volume charge or surface charge). He rejects categorically the idea of a deformable electron. For such an assumption

> implies that owing to the deformation mechanical work would have to be performed and, apart from electromagnetic energy, an internal energy of the electron would have to be accounted for. In this case an electromagnetic interpretation of the theory of cathode or Becquerel rays, purely electric phenomena, would be an impossibility and an electromagnetic foundation of mechanics would have to be renounced from the beginning.

For such an electron Abraham then computes the Lagrangian L, being the difference between the magnetic and the electric energies, and from the Lagrangian he calculates, in exactly the same way as in modern field theory, the momentum. Finally, defining longitudinal mass $\mu_{||}$ as the ratio of the time derivative of the momentum and the acceleration in the direction of motion,

[25] Max Abraham, "Prinzipien der Dynamik des Elektrons," *Annalen der Physik 10,* 105–179 (1903): "Für das Elektron gilt demnach das erste Axiom Newtons in folgender Fassung: War die Bewegung des Elektrons von Anbeginn an eine gleichförmige, rein translatorische, und war die Geschwindigkeit kleiner als die Lichtgeschwindigkeit, so ist, um die Bewegung gleichförmig zu erhalten, keine äussere Kraft oder Drehkraft erforderlich" (p. 142).

and transverse mass $\mu_\perp$ as the corresponding ratio in the direction perpendicular to the motion, Abraham obtains the following result:[26]

$$\mu_\parallel = \frac{q^2}{2ac^2\beta^2}\left(\frac{2}{1-\beta^2} - \frac{1}{\beta}\ln\frac{1+\beta}{1-\beta}\right),$$
$$\mu_\perp = \frac{q^2}{2ac^2\beta^2}\left(\frac{1+\beta^2}{2\beta}\ln\frac{1+\beta}{1-\beta} - 1\right). \tag{21}$$

For small velocities,

$$\mu_\parallel = \mu_\perp = \mu_0 = \frac{2q^2}{3ac^2}, \tag{22}$$

which was Heaviside's result, Eq. (6).

Let us now assume that in addition to the above-mentioned reaction force another external force **K** is acting on the electron and let us also assume that the electron possesses, in addition to the electromagnetic mass μ, an ordinary mechanical mass m (Lorentz's "material mass"). Then the equation of motion reads:

$$\mathbf{K} - \mu\mathbf{w} = m\mathbf{w} \tag{23}$$

or

$$\mathbf{K} = (m + \mu)\mathbf{w} = M\mathbf{w}, \tag{24}$$

where M, as the coefficient of acceleration in the force equation, is the "effective mass" and equal, as we see, to the sum of the mechanical and the electromagnetic masses.

In the general case, when the acceleration is not in the same direction as the motion, Eq. (24) has to be replaced by the more general equation

$$\mathbf{K} = (m + \mu_\parallel)\mathbf{w}_\parallel + (m + \mu_\perp)\mathbf{w}_\perp$$
$$= M_\parallel\mathbf{w}_\parallel + M_\perp\mathbf{w}_\perp, \tag{25}$$

[26] For a detailed derivation of these formulas see also K. Schwarzschild, "Zur Elektrodynamik" (part 3: "Über die Bewegung des Elektrons"), *Göttinger Nachrichten* (1903), pp. 245–278; A. Sommerfeld, "Zur Elektronentheorie," *Göttinger Nachrichten* (1904), pp. 99–130, 363–439, *ibid.* (1905), pp. 201–235. See also H. A. Lorentz, *The theory of electrons* (Dover, New York, 1952), pp. 247–250.

where $\mu_{\parallel}$ is the longitudinal and $\mu_{\perp}$ the transverse electromagnetic mass, $M_{\parallel}$ and $M_{\perp}$ the longitudinal and transverse effective masses, and $\mathbf{w}_{\parallel}$ and $\mathbf{w}_{\perp}$ the components of the acceleration in the directions parallel and perpendicular to the motion.

A purely electromagnetic theory of mass has now, of course, to show that the introduction of a mechanical mass m was unwarranted. In Abraham's view, Kaufmann's experiments, about which we shall speak presently in more detail, verify this conclusion. Abraham's reasoning is as follows. The mechanical mass m, according to Newtonian dynamics, is velocity-independent, while the electromagnetic mass, in virtue of the factor $(1 - v^2/c^2)^{-1/2}$, is velocity-dependent. If, now, the experimental data reveal for the effective mass M the same velocity dependence as for the electromagnetic mass μ, the mechanical mass m is necessarily equal to zero.

In a more accurate way the argument may be stated as follows. From Eq. (21) we know that

$$\mu_{\perp} = \frac{3\mu_0}{4\beta^2}\left(\frac{1 + \beta^2}{2\beta}\ln\frac{1 + \beta}{1 - \beta} - 1\right), \tag{26}$$

that is, $\mu_{\perp}$ is a function of v. From experiments with electrons of two different known velocities v_1 and v_2 the ratio r of the respective effective transverse masses may be found:

$$r = \frac{m + \mu_{\perp}(v_1)}{m + \mu_{\perp}(v_2)}. \tag{27}$$

On the other hand, the ratio s of the two corresponding electromagnetic transverse masses can be computed from Eq. (26):

$$s = \frac{\mu_{\perp}(v_1)}{\mu_{\perp}(v_2)}. \tag{28}$$

Elimination of $\mu_{\perp}(v_2)$ from the last two equations gives

$$\frac{m}{\mu_{\perp}(v_1)} = \frac{s - r}{s(r - 1)}. \tag{29}$$

Thus, if the experimental ratio r coincides, within the margin of error, with the theoretical ratio s, the mechanical mass m may be taken as nil. If, moreover, the electromagnetic mass of a system of charges is equal to the sum of the electromagnetic masses of the individual charges, or, in other words, if the additivity of mass is assured — which Lorentz claimed to be the case provided the fields of the charges may be said not to overlap — then a purely electromagnetic explanation of the kinetic reaction of the electron or of any other particle of a similar constitution seems to have been achieved. Newton's second law of motion would then be a consequence of Maxwell's theory of the electromagnetic field.

Did Kaufmann's famous experiments at the Physical Institute of Göttingen on the deflection of electrons by simultaneous electric and magnetic fields and his determination of e/m really verify Abraham's contention? In the first report on his experiments[27] Kaufmann, summarizing his results, states that the electromagnetic mass μ, which he calls "apparent" (*scheinbare Mass*), is of the same order of magnitude as the mechanical mass m, which he calls "real" (*wirkliche Masse*); with increase of velocity, however, the "apparent mass" exceeds the "real mass" considerably. In view of these results Kaufmann believes that the assumption of a different charge distribution on (or in) the electron may lead to the conclusion that the "real mass" is zero. In a second paper, entitled "On the electromagnetic mass of the electron,"[28] he corrects his previous statements and arrives at the conclusion that the mass of the electron is merely an electromagnetic phenomenon. Meanwhile Abraham, in an article entitled "The dynamics of the electron,"[29] took issue with Kaufmann's terminology: "The often used terms of 'apparent' and

[27] Walter Kaufmann, "Die magnetische und elektrische Ablenkbarkeit der Becquerelstrahlen und die scheinbare Masse der Elektronen," *Göttinger Nachrichten* (1902), pp. 143–155.

[28] Walter Kaufmann, "Über die elektromagnetische Masse des Elektrons," *Göttinger Nachrichten* (1902), pp. 291–296.

[29] Max Abraham, "Die Dynamik des Elektrons," reference 22.

'real' masses may lead to confusion," he warns. "For the 'apparent' mass, in the mechanical sense, is real, and the 'real' mass apparently unreal."[30] On this occasion Abraham also pointed out that, strictly speaking, the electromagnetic mass is not a scalar but a tensor with the symmetry of an ellipsoid of revolution.[31] On the basis of Kaufmann's experiments Abraham concludes his paper with the words: "The inertia of the electron originates in the electromagnetic field."

In the same year, in an address to a scientific meeting in Karlsbad, Abraham announced solemnly: "The mass of the electron is of purely electromagnetic nature."[32]

Hendrik Antoon Lorentz, who hailed this conclusion as "certainly one of the most important results of modern physics,"[33] admitted, however, that "we are free to believe, if we like, that there is some small material mass attached to the electron, say equal to one hundredth part of the electromagnetic one." Although Lorentz thereby acknowledged the inconclusiveness of an experimental verification of Abraham's contention, he nevertheless on the whole seemed to have subscribed to Abraham's view on the basis of the principle of simplicity.

The program of the electromagnetic conception of mass was now fully established: once ponderable atoms and molecules had been reduced to positive and negative charges and their inertial behavior had been explained on the basis of electrodynamics, an extension and generalization of this approach had to be found for molecular and gravitational forces. The whole universe of physics would then amount to merely positive and negative charges and their magnetic fields, all processes in nature would be reduced to convection currents and their radiations, and the

[30] *Ibid.*, p. 24.
[31] *Ibid.*, p. 28.
[32] "Die Masse des Elektrons ist rein elektromagnetischer Art": "Verhandlungen der 74. Naturforscherversammlung in Karlsbad," *Physikalische Zeitschrift 4,* 57 (1902).
[33] H. A. Lorentz, *The theory of electrons* (1909; Dover, New York, ed. 2, 1952), p. 43.

"stuff" of the world would have been stripped of its material substantiality.

The electromagnetic doctrine of mass with its fascinating implications soon began to engage the attention of the learned world. Although it can hardly be said that it ever gained general acceptance, a number of outstanding physicists expressed their approval. Thus, Poincaré, in *Science and method,* declared: "What we call mass would seem to be nothing but an appearance, and all inertia to be of electromagnetic origin."[34] Bucherer, who rechecked Kaufmann's experiments, regarded it as possible "that the mass of the corporeal atoms will eventually reveal itself as merely fictitious."[35] Conway, professor of mathematical physics in University College, Dublin, in a paper entitled "Electromagnetic mass,"[36] developed in terms of quaternions a theory of the electromagnetic mass tensor, the "mass quadric." Comstock[37] and Harkins and Wilson[38] construe a physics of the atom on the basis of the electromagnetic concept of mass.

The early enthusiasm with which the theory was met, however, soon diminished, for it became increasingly clear that the electromagnetic theory of mass was unable to carry out successfully the necessary generalizations for the constituents of matter other than electrons. Furthermore, the experimental verification of the velocity dependence of the electronic mass, which so far was the main evidence for the electromagnetic concept, found a new interpretation in the revolutionary theory of relativity.

For the development of the concept of mass and thus for the

[34] Henri Poincaré, *Science and method* (first French ed., Flammarion, Paris, 1908; Dover, New York, 1952), p. 206.

[35] A. H. Bucherer, *Mathematische Einführung in die Elektronentheorie* (Teubner, Leipzig, 1904), p. 2.

[36] Arthur William Conway, *Selected papers* (Dublin Institute for Advanced Studies, Dublin, 1953), p. 45. The paper appeared originally in the *Scientific Transactions of the Royal Dublin Society* [2], *9,* 51 (1907).

[37] Daniel F. Comstock, "The relation of mass to energy, "*Philosophical Magazine 15,* 1–21 (1908).

[38] William D. Harkins and Ernst D. Wilson, "Wechselseitige Elektromagnetische Masse und die Struktur des Atoms," *Zeitschrift für anorganische und allgemeine Chemie 95,* 1–19 (1916).

development of physical theory in general, the electromagnetic theory of matter was of decisive importance. Till its advent, physicists and philosophers, on the whole, adhered to what was called the substantial concept of physical reality. A physical body, according to this view, is first of all what it is: only on the basis of its intrinsic, invariable, and permanent nature, of which mass was the physical expression and inertial mass the quantitative measure, did it act as it did. The electromagnetic concept, now, proposed to deprive matter of this intrinsic nature, of its substantial mass. Although charge, to some extent at least, fulfills the function of mass, the real field of physical activity is not the bodies but, as Maxwell and Poynting have shown, the surrounding medium. The field is the seat of the energy, and matter ceases to be the capricious dictator of physical events. To interpret mass as quantity of matter, or, more accurately, to regard inertial mass as the measure of quantity of matter, has now lost all meaning. For the primacy of substance has been abandoned. The electromagnetic concept of mass was not only one of the earliest field theories, in the modern sense of the word, but it also fully expressed a fundamental tenet of modern physics and of the modern philosophy of matter: matter does not do what it does because it is what it is, but it is what it is because it does what it does.

CHAPTER **12**

THE RELATIVISTIC CONCEPT OF MASS

The concept of mass from the viewpoint of special relativity shows strikingly how intimately the notion of (inertial) mass is interrelated with the whole structure of physical theory. Since special relativity excludes from its considerations any references to gravitational phenomena, it is clear that only the inertial mass is the subject of the present chapter.

For a profound comprehension of the relativistic concept of mass let us first recall those classical features of our concept that will be affected by relativistic modifications.

In the present context we adopt the elementary (prerelativistic) distinction between conservation (in time) and invariance and covariance with respect to coordinate transformations, despite the fact that conservation laws are ultimately expressions of invariance with respect to certain symmetry operations (the conservation of linear or angular momentum is a consequence of the invariance of the Hamiltonian with respect to a translation or rotation in space, the conservation of energy follows from the invariance under a translation in time, the conservation of charge from the invariance with respect to a gauge transformation, and so on).

We shall say that a quantity Q is "conserved" if the numerical value of Q does not depend on time. We shall call a function F "invariant with respect to a group of transformations T" (briefly,

T-invariant") if the value of F does not change when an arbitrary transformation of T is performed on the arguments of F. Finally, let $P(x, y, z, \ldots)$ be a proposition containing the parameters x, y, z, . . . and let these latter be subjected to an arbitrary transformation belonging to a group of transformations S, so that x becomes x', y becomes y', z becomes z', . . . If the proposition $P'(x', y', z', \ldots)$ has the same logico-mathematical structure in x', y', z', . . . as $P(x, y, z, \ldots)$ had in x, y, z, . . . , we call P' "form invariant" or "covariant" with respect to the group of transformations S, or briefly, "S-covariant." We are fully aware of a certain lack of logical rigor with respect to the present terminology, but for the sake of simplicity we shall not treat the following discussion in a more formalized manner.[1]

Let us start with a few historical remarks. One of the fundamental propositions of classical mechanics is the theorem of conservation of momentum, whose validity follows from the second and third of Newton's laws of motion. Confining our discussion to systems composed of two bodies, we see that from $F_1 = m_1 a_1$ and $F_2 = m_2 a_2$, together with the action-reaction principle $F_1 = -F_2$, we obtain

$$m_1 a_1 + m_2 a_2 = 0,$$

or, after integration,

$$m_1 u_1 + m_2 u_2 = \text{constant},$$

which is the principle of conservation of momentum (m, u, a, and F have the usual signification of mass, velocity, acceleration, and force, respectively).

Newton referred to this principle only *en passant* as the law of conservation of the center of gravity. In corollary 4 to his laws of motion he says:

> The common centre of gravity of two or more bodies does not alter its state of motion or rest by the actions of the bodies among themselves;

[1] For a more rigorous treatment, see J. C. C. McKinsey and P. Suppes, "On the notion of invariance in classical mechanics," *British Journal for the Philosophy of Science* 5, 290–302 (1955).

and therefore the common centre of gravity of all bodies acting upon each other (excluding external actions and impediments) is either at rest, or moves uniformly in a right line.[2]

Although Lagrange, in his *Mécanique analytique*,[3] apparently with reference to this passage, declares Newton to be the discoverer of the principle, it is known that Descartes[4] had already stated it, though in an incomplete formulation, and that Huygens, Mariotte, Wallis, and Wren made use of it in their investigations of the impact of bodies.[5] D'Alembert and, in particular, Lagrange fully realized the fundamental importance of this principle for Newtonian mechanics. So does Mach's definition of mass, as we have seen above.

After this short digression let us return to the prerelativistic concept of mass. Newtonian mechanics asserts the constancy of mass for individual bodies as well as for systems of bodies, and its invariance with respect to Galilean transformations. The fact that an increase or decrease of mass is always interpreted in classical physics as an influx or efflux of matter is, of course, merely a scientific version of the principle of the indestructibility of matter combined with the identification of matter and mass, as we have seen repeatedly.

For future reference and in contradistinction to the relativistic innovations, we shall formally prove a few theorems concerning the prerelativistic concept of mass and its relation to the group of Galilean transformations G:

$$x' = x - vt, \quad y' = y, \quad z' = z, \quad t' = t. \qquad (1)$$

Theorem A: Mass is G-invariant.

Proof: We base our proof on the definition of mass as the

[2] Isaac Newton, *The mathematical principles of natural philosophy*, ed. Florian Cajori (University of California Press, Berkeley, 1947), p. 19.

[3] Louis de Lagrange, *Oeuvres* (Paris, 1888), vol. 1, part 2, p. 259.

[4] René Descartes, *Oeuvres*, ed. Adam and Tannery (Paris, 1907), part 2, sec. 36 of *Principia philosophiae:* "certam tamen et determinatam habet quantitatem, quam facile intelligimus eandem semper in tota rerum universitate esse posse, quamvis in singulis ejus partibus mutetur."

[5] For details see René Dugas, *La mécanique au XVIIe siècle* (Editions du Griffon, Paris, Neuchatel, 1954).

ratio of force to acceleration. The G-invariance of force is a consequence of the fact that Newtonian forces are functions of distance only and distances are G-invariant ($x_2' - x_1' = x_2 - x_1, \ldots$). The G-invariance of acceleration follows immediately from Eqs. (1). Thus, mass, as the ratio of two G-invariant quantities, is itself G-invariant.

Theorem B: The theorem of conservation of momentum is a G-covariant proposition.

Proof: Consider the collision of two bodies with masses m_1 and m_2. Let the velocities of the two bodies before impact be $\mathbf{u}_1$ and $\mathbf{u}_2$, and after impact, $\bar{\mathbf{u}}_1$ and $\bar{\mathbf{u}}_2$, respectively. Relative to the reference system R under discussion, the theorem of conservation of momentum asserts that

$$\sum_{i=1}^{2} m_i \mathbf{u}_i = \sum_{i=1}^{2} m_i \bar{\mathbf{u}}_i. \tag{2}$$

With respect to a reference system R' moving with constant velocity $\mathbf{v}$ relative to the former, the velocities before and after impact are given by the formulas

$$\mathbf{u}_i' = \mathbf{u}_i - \mathbf{v} \quad \text{and} \quad \bar{\mathbf{u}}_i' = \bar{\mathbf{u}}_i - \mathbf{v} \tag{3}$$

(with $i = 1$ or 2) as follows from Eqs. (1). Substituting these values in Eq. (2) we obtain

$$\sum m_i(\mathbf{u}_i' + \mathbf{v}) = \sum m_i(\bar{\mathbf{u}}_i' + \mathbf{v}) \tag{4}$$

or

$$\sum m_i \mathbf{u}_i' = \sum m_i \bar{\mathbf{u}}_i'. \tag{5}$$

Since, according to Theorem A, $m_i = m_i'$, we have

$$\sum m_i' \mathbf{u}_i' = \sum m_i' \bar{\mathbf{u}}_i', \tag{6}$$

which is the conservation theorem of linear momentum for R'.

Theorem C: Mass is velocity-independent.

Proof: Consider the collision of two identical and perfectly inelastic bodies in the reference system R. Let their velocities, before impact, with respect to R, be $\mathbf{u}_1 = \mathbf{u}$ and $\mathbf{u}_2 = -\mathbf{u}$. After impact their velocities are zero. Viewed from R' their velocities before impact are

$$\mathbf{u}_1' = \mathbf{u} - \mathbf{v} \quad \text{and} \quad \mathbf{u}_2' = -\mathbf{u} - \mathbf{v}. \tag{7}$$

The total momentum before impact is

$$m'(u_1')\mathbf{u}_1' + m'(u_2')\mathbf{u}_2', \tag{8}$$

which, by Theorem B, is a vector quantity in the direction of $\mathbf{v}$. Thus

$$m'(u_1')\mathbf{u}_1' + m'(u_2')\mathbf{u}_2' = \alpha\mathbf{v}. \tag{9}$$

In virtue of Eqs. (7) we obtain

$$m'(u_1')(\mathbf{u} - \mathbf{v}) + m'(u_2')(-\mathbf{u} - \mathbf{v}) = \alpha\mathbf{v}. \tag{10}$$

This is generally possible only if

$$m'(u_1') = m'(u_2'), \tag{11}$$

which proves the theorem with respect to the system R'. But in view of Theorem *A* the velocity independence holds in every Galilean system.

After these recapitulations of prerelativistic considerations let us now turn to the concept of mass from the viewpoint of special relativity. In the development of the relativistic notion of mass three different stages may be distinguished, associated with the names of Einstein, Lewis and Tolman, and Minkowski, respectively.

In his historic paper "On the electrodynamics of moving bodies"[6] Einstein developed the notion of a velocity-dependent mass from electrodynamic considerations. Having established in the kinematical part of the paper the so-called Lorentz transformation equations, Einstein considers in the "Electrodynamical part" (part 2) the motion of a particle with charge e and mass m_0, the mass of the particle "as long as its motion is slow." In

[6] Albert Einstein, "Zur Elektrodynamik bewegter Körper," *Annalen der Physik 17*, 891–921 (1905); English translation in *The principle of relativity* (Dover, New York, 1923), pp. 35–65.

the section entitled "Dynamics of the slowly accelerated electron," Einstein arrives at the equations

$$m_0 \frac{d^2x}{dt^2} = eE_x,$$

$$m_0 \frac{d^2y}{dt^2} = eE_y, \qquad (12)$$

$$m_0 \frac{d^2z}{dt^2} = eE_z,$$

which describe the transition of a charged particle from rest to motion with respect to a reference system R. In order to find the law of motion for a particle that initially had the velocity v with respect to R, Einstein considers the situation from the point of view of a system R' that is moving with velocity v relative to R. In this system R' the situation is identical with the former case of the transition from rest to motion (in R). According to the principle of relativity,

$$m_0 \frac{d^2x'}{dt'^2} = eE'_x,$$

$$m_0 \frac{d^2y'}{dt'^2} = eE'_y, \qquad (13)$$

$$m_0 \frac{d^2z'}{dt'^2} = eE'_z.$$

The mass of the particle, as viewed from R', is again that of a slowly moving particle and by the principle of relativity must be equal to m_0. With the help of the Lorentz equations the preceding equations can now be transformed into the coordinates of R, yielding, for a motion along Ox,

$$m_0\gamma^3 \frac{d^2x}{dt^2} = eE'_x,$$

$$m_0\gamma^2 \frac{d^2y}{dt^2} = eE'_y, \qquad (14)$$

$$m_0\gamma^2 \frac{d^2z}{dt^2} = eE'_z,$$

where γ is equal to $(1 - v^2/c^2)^{-1/2}$. Since the right-hand members of Eqs. (14) "are the components of the ponderomotive force acting upon the electron, and are so indeed as viewed in a system moving at the moment with the electron, with the same velocity as the electron," a comparison[7] with the traditional formula mass $\times$ acceleration $=$ force shows that the longitudinal mass is equal to $\gamma^3 m_0$ and the transverse mass equal to $\gamma^2 m_0$:

$$\text{Longitudinal mass} = \frac{m_0}{(1 - v^2/c^2)^{3/2}}, \tag{15}$$

$$\text{Transverse mass} = \frac{m_0}{(1 - v^2/c^2)^{1/2}}. \tag{16}$$

Einstein's remark that "with a different definition of force and acceleration we should naturally obtain other values for the masses"[8] shows clearly that he recognized the arbitrariness in his definition of mass. In fact, if the forces were defined so that the laws of momentum and energy assume the simplest form, then in view of the identity

$$\gamma^3 \frac{d^2x}{dt^2} = \frac{d}{dt}\left(\gamma \frac{dx}{dt}\right),$$

Eq. (14) would assume the form

$$\frac{d}{dt}\left[\frac{m_0 v}{(1 - v^2/c^2)^{1/2}}\right] = \text{force vector.} \tag{17}$$

Comparing this expression with Newton's original definition of force as the rate of change of momentum, and defining mass as the coefficient of velocity in the expression for the momentum, we see that the velocity dependence of mass is given by the general formula

$$m = \frac{m_0}{(1 - v^2/c^2)^{1/2}}. \tag{18}$$

That a result of such vital importance for mechanics proper

[7] In connection with Einstein's present definition of force see Max Planck, "Das Prinzip der Relativität und die Grundgleichungen der Mechanik," *Berichte der Deutschen Physikalischen Gesellschaft* 4, 136–141 (1906).

[8] Reference 6 (Dover ed.), p. 63.

was obtained from conclusions based on the Maxwell-Hertz theory of electromagnetism was generally felt as a serious shortcoming of the theory of relativity. It was thus only natural to search for a derivation of Eq. (18) without recourse to non-mechanical theories. The accomplishment of this task is the second stage in the development of the relativistic conception of mass.

In a paper entitled "The principle of relativity and non-Newtonian mechanics,"[9] Lewis and Tolman published in 1909 a derivation of Eq. (18), based on the theorem of conservation of momentum and on the Lorentz transformation equations. They consider two systems of reference a and b, moving uniformly in the direction of the common x-axis.

An experimenter A on the first system constructs a ball of some rigid material, with a volume of one cubic centimeter, and sets it in motion, with a velocity of one cm per sec, towards the system b (in a direction perpendicular to the line of relative motion of the two systems). On the other system, an experimenter B constructs of the same material a similar ball with a volume of one cubic centimeter in his units, and imparts to it, also in his units, a velocity of one cm per sec towards a. The experiment is so planned that the balls will collide and rebound over their original paths. Since the two systems are entirely symmetrical, it is evident by the principle of relativity, that the (algebraic) change in velocity of the first ball, as measured by A, is the same as the change in velocity of the other ball, as measured by B. This being the case, the observer A, considering himself at rest, concludes that the real change in velocity of the ball b is different from that of his own, for he remembers that while the unit of length is the same in this transverse direction in both systems, the unit of time is longer in the moving system.

Velocity is measured in centimeters per second, and since the second is longer in the moving system, while the centimeter in the direction which we are considering is the same in both systems, the observer A, always using the units of his own system, concludes that the change in velocity of the ball b is smaller in the ratio $(1 - \beta^2)^{1/2}: 1$ than the change in velocity of the ball a. The change in velocity of each ball multiplied by its mass gives its change in momentum. Now, from the law of conservation of momentum, A assumes that each ball experiences the same change in momentum, and therefore since he has already decided that the ball b has experienced a smaller change of velocity in the ratio

[9] Gilbert N. Lewis and Richard C. Tolman, *Philosophical Magazine 18*, 510–523 (1909).

$(1-\beta^2)^{1/2}:1$ he must conclude that the mass of the ball in system b is greater than that of his own in the ratio $1:(1-\beta^2)^{1/2}$. In general, therefore, he must assume that the mass of a body increases with its velocity.[10]

Once more we see that mass is defined so as to satisfy a certain mathematical equation. In the present case this equation is the law of conservation of momentum whose G-invariance is tacitly assumed by Lewis and Tolman.

Campbell, in his paper on "Relativity and the conservation of momentum," criticizes the reasoning in the above-quoted passage. In particular, he directs his attack against the vagueness of expressions such as "the real change in velocity" used by the two authors. "When he (Tolman) proceeds to calculate the 'real change,'" argues Campbell, "without explaining what he means by the term, I have the suspicion that he is using words to which he cannot attach any significance whatsoever."[11] Although this insistence on greater accuracy in expression is certainly justified, Campbell was undoubtedly wrong in rejecting the result of these investigations. In particular, his contention that the result would not be valid for collision in the line of relative motion was soon disproved in another paper by Tolman in 1912,[12] in which he derived Eq. (18) for the case of a head-on collision in the common direction of the motion of the two reference systems.

Meanwhile Epstein derived the formula in a rigorous way for impact perpendicular to the direction of motion.[13] The general case of elastic collisions of two moving bodies on the basis of purely mechanical considerations was the subject of a detailed investigation by Ferencz Jüttner.[14]

[10] *Ibid.*, p. 517.

[11] Norman Campbell, *Philosophical Magazine 21,* 626–630 (1911).

[12] Richard C. Tolman, "Non-Newtonian mechanics, the mass of a moving body," *Philosophical Magazine 23,* 375–380 (1912).

[13] Paul S. Epstein, "Über relativistische Statik," *Annalen der Physik 36,* 779–795 (1911). Epstein's derivation of formula (18) is given in section 4 of the paper (Appendix: Impulssatz und Massentransformation), p. 792. For a modern presentation of Epstein's derivation see P. G. Bergmann, *Introduction to the theory of relativity* (Prentice-Hall, New York, 1950), pp. 87–91.

[14] Ferencz Jüttner, "Die Gesetze des Stosses in der Lorentz-Einsteinschen Relativtheorie," *Zeitschrift für Mathematik und Physik 62,* 410–433 (1913).

All these investigations assume the validity of the principle of conservation of momentum and by cleverly designed thought experiments derive the velocity dependence of mass. Their common point of departure is thus the following theorem, which should be compared with Theorems *A–C*:

Theorem D: If the theorem of conservation of momentum is an L-covariant proposition, mass cannot be velocity-independent.

Proof: Consider a completely inelastic collision of two identical bodies moving in the direction of the x-axis of a reference system R with equal and opposite velocities $u_1 = u$ and $u_2 = -u$. After impact their velocities are zero. With respect to a reference system R', moving in the direction of the common x-axis relative to R with velocity v, the velocities of the two bodies are, before impact,

$$u_1' = \frac{u - v}{1 - uv/c^2} \quad \text{and} \quad u_2' = \frac{-u - v}{1 + uv/c^2}, \tag{19}$$

and after impact,

$$\bar{u}_1' = \bar{u}_2' = -v, \tag{20}$$

as follows easily from the Lorentz transformation equations. Let us assume that mass is velocity-independent. From this assumption and the L-covariant conservation law we obtain for R'

$$m'u_1' + m'u_2' = m'\bar{u}_1' + m'\bar{u}_2'. \tag{21}$$

Substituting from Eqs. (19) and (20), and after canceling by the velocity-independent mass m', we have

$$\frac{u - v}{1 - uv/c^2} + \frac{-u - v}{1 + uv/c^2} = -2v. \tag{22}$$

Choosing $v = u$, we obtain

$$\frac{2u}{1 + u^2/c^2} = 2u, \tag{23}$$

which shows that our assumption of velocity independence of mass is inconsistent with the L-covariance of the conservation theorem.

Relativity thus had the choice either of abandoning the Lorentz covariance of the conservation theorem of linear momentum or of accepting the conclusion that mass is a velocity-dependent quantity. The second alternative proved of greater methodological convenience.

The development so far did not utilize the concept of four-vectors by means of which the theory of relativity, and relativistic dynamics in particular, can be given a much neater and philosophically more satisfying representation, as we know since Minkowski's epoch-making publication of 1908.[15] In the language of four-vectors — and this is the third of the three stages mentioned above — the dynamical properties of a particle are characterized by the so-called energy-momentum vector P^i which is postulated to be (1) always parallel to the four-velocity U^i ($= dx/d\tau,\ dy/d\tau,\ dz/d\tau,\ d(ict)/d\tau$, where $d\tau$ is the element of proper-time) and (2) constant in time for a free particle. These conditions imply that

$$P^i = m_0 U^i, \tag{24}$$

where m_0 is an invariant to be called the proper mass (or rest mass) of the particle. In virtue of the relation $d\tau = dt\ (1 - \beta^2)^{1/2}$, the spatial components of P^i are

$$P^1 = \frac{m_0\, dx/dt}{(1-\beta^2)^{1/2}}, \quad P^2 = \frac{m_0\, dy/dt}{(1-\beta^2)^{1/2}}, \quad P^3 = \frac{m_0\, dz/dt}{(1-\beta^2)^{1/2}}, \tag{25}$$

or, in three-dimensional notation,

$$\mathbf{P} = \frac{m_0}{(1-\beta^2)^{1/2}}\mathbf{v}. \tag{26}$$

Identifying — as in classical mechanics — the coefficient of the velocity in the expression for the momentum as the "mass" m of the particle, we obtain

[15] Hermann Minkowski, "Space and time" (an address delivered at the 80th Assembly of German Scientists and Physicians at Cologne, 21 September 1908); English translation in *The principle of relativity* (Dover, New York, 1923), pp. 73–91. See also "Die Grundgleichungen für die elektromagnetischen Vorgänge in bewegten Körpern," *Göttinger Nachrichten* (1908), pp. 53–111; reprinted in *Mathematische Annalen 68*, 472–499 (1910).

$$m = \frac{m_0}{(1 - \beta^2)^{1/2}}. \tag{27}$$

A formal proof of the conservation (constancy in time) of the proper mass of a free particle (without interaction) can easily be given. Since

$$\frac{dP^i}{d\tau} = \frac{dm_0}{d\tau} U^i + m_0 \frac{dU^i}{d\tau} = 0, \tag{28}$$

multiplication by U^i gives

$$\frac{dm_0}{d\tau} U^i U^i + m_0 \frac{dU^i}{d\tau} U^i = 0. \tag{29}$$

The second term in the left-hand member of this equation vanishes because of the constant length of the four-vector. Consequently,

$$\frac{dm_0}{d\tau} = 0, \tag{30}$$

or, since $d\tau = dt\ (1 - \beta^2)^{1/2}$,

$$\frac{dm_0}{dt} = 0. \tag{31}$$

The derivation of Eq. (27) with the help of the calculus of four-vectors exposes, as did the former considerations concerning collisions of particles, the conceptual or definitional character of velocity-dependent mass. It is the new relation between space and time, in other words, the Lorentz-Minkowski kinematics, that produces the peculiar functional dependence of momentum on velocity and consequently the velocity dependence of mass. It is not a new property of matter that has been discovered nor a mysterious or secret trait of nature that has been unmasked by science.

And yet, was not the relativistic Eq. (27) subjected to experimental verification? In fact, as Abraham's electromagnetic theory of mass led to a functional dependence of mass that differs from the relativistic velocity dependence, much experimental work has been done since 1906 to force a decision between the rival theories.

As we have seen,[16] the velocity dependence of the transverse mass is given in the electromagnetic theory of matter by the formula

$$m = f_1(\beta),$$

where

$$f_1(\beta) = \frac{3}{4}\frac{m_0}{\beta^2}\left(\frac{1+\beta^2}{2\beta}\ln\frac{1+\beta}{1-\beta} - 1\right);$$

in the theory of relativity, on the other hand,

$$m = f_2(\beta),$$

where

$$f_2(\beta) = m_0(1 - \beta^2)^{-1/2}.$$

Expanding $f_1(\beta)$ and $f_2(\beta)$ into power series in β,

$$f_i(\beta) = f_i(0)(1 + a_{i1}\beta + a_{i2}\beta^2 + \cdots) \quad (i = 1, 2),$$

we can easily see that in Abraham's theory $a_{11} = 2/5$, $a_{12} = 9/35$, while in Einstein's theory $a_{21} = \frac{1}{2}$ and $a_{22} = 3/8$. Measurements of the charge-to-mass ratio e/m as a function of $v = \beta c$ should therefore make a decision possible.

Kaufmann's early experiments were at first, as we have seen, enthusiastically interpreted as experimental verifications of Abraham's concept of mass. Thus, for example, K. Schwarzschild begins his article "On electrodynamics"[17] with the statement: "Mr. Abraham's theory has in a most fascinating manner been confirmed by the experiments performed by Mr. Kaufmann." But as early as 1908 various experiments, and in particular those carried out by A. H. Bucherer,[18] seemed to supply evidence in favor of

[16] Page 148.

[17] K. Schwarzschild, "Zur Elektrodynamik," *Göttinger Nachrichten* (1903), 245–278. The statement reads: "Herr Abrahams Theorie hat sich an den Versuchen von Herrn Kaufmann in der erstaunlichsten Weise bewährt."

[18] A. H. Bucherer, "Messungen an Becquerelstrahlen. Die experimentelle Bestätigung der Lorentz-Einsteinschen Theorie," *Physikalische Zeitschrift 9*, 755–762 (1908); "Die experimentelle Bestätigung des Relativitätsprinzips," *Annalen der Physik 28*, 513–536 (1909); "Nachtrag zu meiner Arbeit 'Bestätigung des Relativitätsprinzips,'" *Annalen der Physik 29*, 1063 (1909); A. Bestelmeyer, "Bemerkungen zu der Abhandlung Herrn A. H. Bucherers 'Die experi-

the relativistic conception. The velocity dependence of the electron mass became a subject of numerous experiments and more numerous controversies.[19] Among the more important investigations of this kind those of Weiss and Cotton,[20] Classen,[21] Wolz,[22] Gmelin,[23] Hupka,[24] Guye and Ratnovsky,[25] Malassez,[26] Schäfer,[27] and Neumann[28] have to be mentioned. Similarly, the measurements of the charge-to-mass ratio of electrostatically accelerated electrons carried out in 1921 by Guye and Lavanchy,[29] the experiments in nuclear spectroscopy performed in 1940 by Rogers, McReynolds, and Rogers,[30] and the determinations of the charge-

mentelle Bestätigung des Relativitätsprinzips,'" *Annalen der Physik 30,* 166–174 (1909); A. H. Bucherer, "Antwort auf die Kritik des Herrn A. Bestelmeyer bezüglich meiner experimentellen Bestätigung des Relativitätsprinzips," *Annalen der Physik 30,* 974–986 (1909); A. Bestelmeyer, "Erwiderung auf die Antwort des Herrn A. H. Bucherer," *Annalen der Physik 32,* 231–235 (1910).

[19] Of which the exchange between Bucherer and Bestelmeyer (see reference 18) is only one example. More recently a detailed analysis of Bucherer's method has been carried out by C. T. Zahn and A. A. Spees in their paper, "A critical analysis of the classical experiments on the relativistic variation of electron mass," *Physical Review 53,* 511–521 (1938). The authors come to the conclusion that the velocity filters employed by Bucherer were deficient so that the resolution for velocities above 0.7*c* was too low for a decisive result.

[20] *Journal de Physique 6,* 429 (1907).

[21] *Verhandlungen der Deutschen Physikalischen Gesellschaft 10,* 700 (1908).

[22] K. Wolz, "Die Bestimmung von e/m," *Annalen der Physik 30,* 273–288 (1909).

[23] P. Gmelin, "Der Zeemaneffekt einiger Quecksilberlinien in schwachen Magnetfeldern," *Annalen der Physik 28,* 1079–1087 (1909).

[24] E. Hupka, "Beitrag zur Kenntnis der trägen Masse bewegter Elektronen," *Annalen der Physik 31,* 169–204 (1910).

[25] C. E. Guye and S. Ratnovsky, "Sur la variation de l'inertie de l'electron en fonction de la vitesse dans les rayons cathodiques et sur le principe de relativité," *Comptes rendus 150,* 326–329 (1910). On p. 329 the authors declare: "Le principe de relativité se trouve en accord avec l'expérience."

[26] *Annales de Chimie et de Physique 23,* 231 (1911).

[27] C. Schäfer, "Die träge Masse schnell bewegter Elektronen," *Physikalische Zeitschrift 14,* 1117–1118 (1913).

[28] G. Neumann, "Die träge Masse schnell bewegter Elektronen," *Annalen der Physik 45,* 529–579 (1914). Summing up the situation in 1914 Neumann concludes that for $0.4 \leqslant \beta \leqslant 0.7$ the deflection methods speak in favor of the relativistic variation, but for $0.7 < \beta \leqslant 0.8$ no certain conclusion can be drawn.

[29] C. E. Guye and C. Lavanchy, *Mémoires de la Société de Physique de Genève 39,* 315 (1921).

[30] M. M. Rogers, A. W. McReynolds, and F. T. Rogers, Jr., "A determination of the masses and velocities of three radium B beta-particles," *Physical Review 57,* 379–383 (1940).

to-mass ratio of protons which Grove and Fox performed with the 140-inch synchrocyclotron of the Carnegie Institute of Technology in 1953 — three of the most important experiments of this kind in the course of the past 40 years — were generally interpreted as settling the answer in favor of the relativistic formula. However, Faragó and Jánossy,[31] who recently reexamined the whole experimental evidence, came to the conclusion that the experiments carried out so far "support the validity of the relativistic formula far less than is usually supposed to be the case." In fact, in their view, only the study of the fine structure of hydrogenlike spectra unequivocally confirms the relativistic equation with high precision.

Since experiments with free electrons do not seem so far to give a clear-cut decision between the formula of Abraham and that of Einstein and since the spectroscopic evidence — in addition to being merely indirect — covers only a rather narrow range of velocities, an unambiguous direct verification of this all-important velocity dependence of mass is certainly still a matter of serious concern for experimental physics. The general consensus is, however, that the problem is only that of increased precision in technique.

How, then, is it possible that the definitional or conventional postulation of the velocity dependence of mass became a factual, empirically ascertainable, and experimentally measurable situation?

This specific problem was recently discussed by Karl Vogtherr in an article entitled "The variability of mass in the theory of relativity.[32] Vogtherr confronts the positivistic physicist, who claims that "the variability of mass is as little ascertainable by means of measurement as anything else, if measurement means

[31] P. S. Faragó and L. Jánossy, "Review of the experimental evidence for the law of variation of the electron mass with velocity," *Nuovo cimento 5,* 1411–1436 (1957).

[32] Karl Vogtherr, "The variability of mass in the theory of relativity," *Methodos 9,* 199–207 (1957).

the nonarbitrary apprehension of relationships in an external world," with the realistic physicist, according to whom "the variability of mass is a measurable relationship, and as such it is ascertainable and satisfies the formula of relativity theory." The present author does not accept Vogtherr's conclusion that "the variability of mass is, according to the theory of relativity, not unique, because v and the body of reference which belongs thereto are not uniquely ascertainable" and thus an experimental ascertainability would lead to a *contradictio in adjecto*. "For a reality which is ascertainable as ambiguous, which is 'so' and at the same time 'other,' is unthinkable."[33] The answer to the problem, it seems, is this. As much as the prerelativistic definition of the concept of inertial mass was part and parcel of the physical theory as a whole, so also is the experimental determination of relativistic mass most intimately interrelated with a body of numerous operations and interpretations and not to be considered as an isolated fact.

To make this point clear, let us discuss "the experimental verification of the relativistic velocity dependence of mass" in greater detail. In these classical experiments high-speed electrons with charge e, emitted by radioactive sources or artificially accelerated, are subjected to the influence of electric and magnetic fields, E and H. First, by a critical balancing of the beam between the condenser plates where the electric force eE cancels the magnetic force evH, the velocity v is determined as E/H. Secondly, by measuring the radius of curvature r of the electron beam outside the condenser the ratio $e/m(v)$ is determined from the equation

$$evH = m(v)v^2/r.$$

Thus a correlation is established between $m(v)$ and v, which — if in conformity with Eq. (27) — is said "to verify the velocity

[33] *Ibid.*, p. 203.

dependence of mass." However, the equation above could equally well be written

$$ev(1 - v^2/c^2)^{1/2}H = m_0v^2/r$$

without ever mentioning the idea of a "variable mass." The deflection experiments would then test a relatively complicated relation between r and v, taking e and m_0 as constants. Moreover, the results of the deflection experiments can also be interpreted as showing a velocity dependence of the charge e, as indeed Bush in his paper "The force between moving charges" [34] suggested. That still other interpretations of these empirical results are possible has been shown also by O'Rahilly.[35]

We thus realize that the ambiguity in the interpretation of empirical data is the result of the definitory character of the concepts under discussion. As soon, however, as specific, albeit arbitrary, definitions for the theoretical concepts are adopted, the process of verification — within the confines of the same theory — loses its ambiguity and becomes a uniquely meaningful operation. The definitional character of the underlying concepts has its analogue in the interpretation of the experimental results. The moment of arbitrariness operative in the conceptual construction of the theory reappears in the descriptive interpretation of the empirical data. "Mass," in relativity, is merely the result of certain operations, the definitions or specifications of which are intimately connected with spatio-temporal considerations. Only in virtue of these connections does the outcome of the measuring operations depend on the velocity.

With increase of velocity, mass increases. It is thus also obvious that all associations with the historical predecessor of mass, the *quantitas materiae*, are completely severed, as they were in the

[34] V. Bush, "The force between moving charges," *Journal of Mathematics and Physics 5,* 129 (1925–26).

[35] Alfred O'Rahilly, *Electromagnetics, a discussion of fundamentals* (Cork University Press, Cork; Longmans, Green, London, New York, 1938). See also Parry Moon and Domina Eberle Spencer, "The new electrodynamics and its bearing on relativity," in *Kritik und Fortbildung der Relativitätstheorie,* Karl Sapper, ed. (Akademische Druck- und Verlagsanstalt, Graz, 1958), pp. 144–159.

electromagnetic theory of matter. Otherwise one would have to draw the conclusion that motion creates matter, a result against which Bullialdus, in 1639, vehemently protested.[36]

[36] Ismael Bullialdus (Boulliaud), *Philolai sive dissertationis de vero systemate mundi libri quatuor* (Amsterdam, 1639). In his investigation of the acceleration (and gravity) of free fall (book 1, chap. 4), Bullialdus refutes the theory according to which a body, during and through its descending motion, acquires additional gravity. If this were so, he says, "it would follow that something like matter can be produced by local motion; this, however, is erroneous, because with the position of local motion is posited only the *ubi* and no new substance can result therefrom."

CHAPTER 13

MASS AND ENERGY

Since Maxwell's theory of electromagnetism, from the historical point of view, was the earliest scheme of a field theory to reach a high level of consistent elaboration and logical maturity, it is not surprising that the modern development of the concept of mass is intimately related to various phases of this theory. Not only did the electromagnetic concept of mass, as the name implies, develop from this theory but, as we have seen in the preceding chapter, the relativistic concept of mass — both historically and intrinsically[1] — also originated therefrom. Maxwell's theory of the electromagnetic field also brought forth, though in a less direct manner, another concept of mass which has been called "one of the greatest discoveries of the twentieth century."[2]

Maxwell showed that the work necessary to build up an electromagnetic field can be regarded as equivalent to the energy produced in space with a density w given by the formula

$$w = \frac{1}{8\pi}(E^2 + H^2), \tag{1}$$

[1] Intrinsically, because the Lorentz equations, basic for the derivation of the velocity dependence of mass, describe those transformations with respect to which Maxwell's equations are covariant.

[2] Sir Edmund Whittaker, *History of the theories of aether and electricity* (Nelson, Glasgow, London, 1953), vol. 2, p. 51.

where **E** is the electric and **H** the magnetic field vector. This formula applies also to nonstatic fields where, in conformity with the energy-conservation theorem, energy has to flow from one place to another to compensate for changes that occur in a particular region of space. The energy transported in unit time across unit area is given by the vector

$$\mathbf{S} = \frac{c}{4\pi} \mathbf{E} \times \mathbf{H},$$

as John Henry Poynting showed in 1884.[3] A transport of energy, it is true, had already before Poynting been associated with electric currents, but the electric energy was regarded as confined to the conducting wires along which it was conveyed. In Poynting's theory of energy flow, however, the transport of energy is no longer confined to conductors. The surrounding medium or empty space is the arena where energy moves. Energy, thus disjoined from matter, raised its ontological status from a mere accident of a mechanical or physical system to the autonomous rank of independent existence, and matter ceased to be the indispensable vehicle for its transport. As a result of this emancipatory change or reification of energy, the idea that only differences of energy are of physical significance had to be abandoned and an absolute magnitude had to be ascribed to energy, as we shall see later on. In contrast to the older theories of action-at-a-distance, according to which energy disappears at a point A in space and reappears at a point C without affecting an intermediate point B, the new field theory of electromagnetism endows energy with a continuous existence in time as well as in space. Lodge, commenting on Poynting's ideas, expresses this fact as follows:

On the new plan we may label a bit of energy and trace its motion and change of form, just as we may ticket a piece of matter so as to identify it in other places under other conditions; and the route of the energy may be discussed with the same certainty that its existence was continuous as would be felt in discussing the route of some lost luggage

[3] *Philosophical Transactions 175*, 343 (1884).

which has turned up at a distant station in however battered and transformed a condition.[4]

This view of the nature of energy was, however, not yet generally accepted. Hertz, for example, in his "Investigations on the propagation of electrical force"[5] still questions the physical significance of speaking of a localization of energy or of the continuity in its propagation. "For investigations of this kind," he claims, "had not yet been carried out even with respect to simple energy transformations in ordinary mechanics itself." Yet admitting such possibility he continues that "the conceptual obscurities involved in such considerations have still to be clarified."[6] Georg Helm supports Hertz's demand for further investigation, if one regards the transport of energy as a real physical principle; if, however, such transport is conceived merely as an analogy or as a verbal description of the equation

$$\frac{\partial w}{\partial t} = -\operatorname{div} S, \tag{2}$$

no conceptual difficulties are encountered. Helm rejects the first alternative, for to ascribe to energy substantial existence seems to him "a dubious aberration from the lucid intelligibility of Robert Mayer's original conceptions. Nothing is absolute, only relations are accessible to science."[7]

The program that Hertz had suggested was carried out, at least for the mechanics of elastically deformable bodies, by Gustav Mie in 1898 in his "Outline of a general theory of energy transfer."[8] In this publication, whose importance for the development

[4] Oliver Lodge, "On the identity of energy in connection with Mr. Poynting's paper on the transfer of energy in an electromagnetic field; and on the fundamental forms of energy," *Philosophical Magazine 19,* 482 (1885).

[5] Heinrich Hertz, "Untersuchungen über die Ausbreitung der elektrischen Kraft," *Gesammelte Werke* (Leipzig, 1894), vol. 2, p. 234.

[6] *Ibid.,* p. 293.

[7] "Es existiert kein Absolutes, nur Beziehungen sind unserer Erkenntnis zugänglich." Georg Helm, *Die Energetik nach ihrer geschichtlichen Entwicklung* (Leipzig, 1898), p. 362.

[8] Gustav Mie, "Entwurf einer allgemeinen Theorie der Energieübertragung, *Wiener Sitzungsberichte 107,* pp. 1113–1182 (1898).

of modern physics can hardly be overestimated, Mie showed that not only a fluid, moving under pressure, conveys by transport energy and conducts energy in proportion to its pressure and velocity, but also every deformation of an elastically stressed body gives rise to an energy flux of a precisely determinable amount. Thus, in a driving belt, operating between a motor and an energy-consuming machine, the energy flux, pointing in a direction opposite to the motion of the stretched part of the belt, is a precisely mathematizable quantity. Showing the universal validity of the equation of continuity (2), Mie declares that energy can be visualized as a fluid spread out in space with density w and flowing with a current density **S**. "All energy changes are consequences of real energy currents." [9] Mie, however, was not yet in a position to draw the final conclusion. For him, matter and energy were still different aspects of physical reality. The similarity between his results concerning the flow of energy and the hydrodynamic equation of continuity for ponderable matter or fluids was in his view only accidental.[10]

> The velocity of matter in motion can be ascertained by inertial effects, such as impact phenomena, centrifugal forces, etc. These effects make it possible to measure the velocity of a fluid in motion, for instance, without taking recourse to a measurement of the current intensity. Nobody, however, has ever spoken of the inertia of energy in motion.[11]

Only two years later Poincaré published a paper entitled "The theory of Lorentz and the principle of reaction,"[12] in which he characterized electromagnetic energy as "a fluid endowed with inertia." He first shows that electromagnetic radiation possesses momentum of an amount equal to the Poynting vector divided by the square of the velocity of light:

[9] *Ibid.*, p. 1129. For a precise definition of Mie's concept of "wirklicher Energiestrom," see the introductory sections of his paper.

[10] "Ich bemerke dabei ausdrücklich, dass die Ähnlichkeit zwischen diesem Satz und dem von der Continuität der Masse nur eine ganz äusserliche ist." *Ibid.*

[11] "Niemand hat aber jemals von einer Trägheit bewegter Energie gesprochen." *Ibid.*, p. 1121.

[12] Henri Poincaré, "La théorie de Lorentz et le principe de réaction," *Archives Néerlandaises des sciences exactes et naturelles 2*, 232 (1900); reprinted in *Recueil de travaux offerts par les auteurs à H. A. Lorentz* (The Hague, 1900).

$$\mathbf{g} = \mathbf{S}/c^2. \tag{3}$$

Then, putting $\mathbf{S}$ equal to $E\mathbf{c}$, where E is the electromagnetic energy absorbed by a body of mass m, he applies the theorem of conservation of momentum in order to compute the recoil velocity $\mathbf{v}$ of the absorbing body by means of the equation

$$m\mathbf{v} = \mathbf{S}/c^2 = (E/c^2)\mathbf{c}. \tag{4}$$

Thus, the mass or inertia of electromagnetic radiation is, by dimensional analysis of Eq. (4), equal to E/c^2. Poincaré, however, still rejects the idea that this "mass" possesses the character of an indestructible substance. He says:

> We may regard electromagnetic energy as a fictitious fluid which has a density w and which is displaced in space in conformity with Poynting's laws. One has, however, to admit that this fluid is not indestructible.[13]

Herbert E. Ives recently presented in an article entitled "Derivation of the mass-energy relation"[14] a detailed reconstruction of Poincaré's paper in the light of Poincaré's "principle of relativity"[15] and showed that Poincaré's arguments, if only pursued to their final conclusions, lead necessarily to the following relation between electromagnetic energy and inertia:

$$m = E/c^2, \tag{5}$$

where m is the change of inertial mass and E the energy involved (absorbed or emitted).

In 1904 Hasenöhrl[16] showed that electromagnetic energy E, enclosed in an empty box with perfectly reflecting walls, behaves,

[13] "Nous pouvons regarder l'énergie électromagnétique comme un fluide fictif dont la densité est w et qui se déplace dans l'espace conformément aux lois de Poynting. Seulement il faut admettre que ce fluide n'est pas indestructible." *Ibid.*, p. 256.

[14] *Journal of the Optical Society of America 42*, 540–543 (1952).

[15] See Henri Poincaré, "L'état actuel et l'avenir de la physique mathématique," *Congress of Arts and Sciences* (St. Louis, September 24, 1904); reprinted in *La Revue des Idées* (November 15, 1904).

[16] Fritz Hasenöhrl, "Zur Theorie der Strahlung in bewegten Körpern," *Annalen der Physik 15*, 344–370 (1904); *Wiener Sitzungsberichte 113*, 1039 (1904); "Zur Theorie der Strahlung in bewegten Körpern (Berichtigung)," *Annalen der Physik 16*, 589–592 (1905).

when the box is set in motion, as if it had a mass proportional to E.[17]

It is generally said that "the theorem of the inertia of energy in its full generality was first stated by Einstein (1905)."[18] The article referred to is Einstein's paper, "Does the inertia of a body depend upon its energy content?"[19] On the basis of the Maxwell-Hertz equations of the electromagnetic field Einstein contended that "if a body gives off the energy E in the form of radiation, its mass diminishes by E/c^2."[20] Generalizing this result for all energy transformations Einstein concludes: "The mass of a body is a measure of its energy content."[21]

It is a curious incident in the history of scientific thought that Einstein's own derivation of the formula $E = mc^2$, as published in his article in the *Annalen der Physik,* was basically fallacious. In fact, what for the layman is known as "the most famous mathematical formula ever projected" in science[22] was but the result of a *petitio principii,* the conclusion of begging the question. This statement, of course, does not diminish in the least the importance of Einstein's contribution to the problem, since the mass-energy relation is a necessary consequence of the theory of relativity and can be deduced from the fundamental assumptions of the theory by various methods, and not only by the method employed by Einstein in his original publication. The logical illegitimacy of Einstein's derivation has been shown by Ives.[23]

Einstein's reasoning is based on a comparison of the total and the kinetic energy of a body and the energy content of the emitted radiation in two reference systems, S and S'. In S the

[17] In the 1904 papers the coefficient in question is $8/3c^2$; in the 1905 paper it is $4/3c^2$.

[18] Max Born, *Atomic physics* (Blackie, London, Glasgow, ed. 6, 1957), p. 55.

[19] Albert Einstein, "Ist die Trägheit eines Körpers von seinem Energieinhalt abhängig?" *Annalen der Physik 18,* 639–641 (1905); English translation in *The principle of relativity* (Dover, New York, 1923), pp. 67–71.

[20] *Ibid.,* p. 71.

[21] *Ibid.:* "Die Masse eines Körpers ist ein Mass für dessen Energieinhalt" (p. 641 of the original paper).

[22] William Cahn, *Einstein* (Citadel Press, New York, 1955), p. 26.

[23] See reference 14.

body is at rest; with respect to S' the body moves with a velocity v. If E_0 and E_0' are the total energy of the body before the emission, relative to S and S', respectively, and E_1 and E_1' the corresponding energy after the emission, T_0' and T_1' the kinetic energy of the body with respect to S', before and after the emission, and E the energy content of the radiation as judged from S, Einstein proves correctly that

$$(E_0' - E_0) - (E_1' - E_1) = E\left[\frac{1}{(1 - v^2/c^2)^{1/2}} - 1\right]. \tag{6}$$

If, further, m_0 and m_1 denote the mass of the body relative to S, before and after the emission,

$$T_0' = m_0c^2\left[\frac{1}{(1 - v^2/c^2)^{1/2}} - 1\right] \tag{7}$$

and

$$T_1' = m_1c^2\left[\frac{1}{(1 - v^2/c^2)^{1/2}} - 1\right]. \tag{8}$$

Einstein now mistakenly put $E_0' - E_0$ equal to $T_0' + C$ (C is a constant) and $E_1' - E_0$ equal to $T_1' + C$, and thus obtained by subtraction, and in virtue of Eq. (6),

$$T_0' - T_1' = E\left[\frac{1}{(1 - v^2/c^2)^{1/2}} - 1\right] \tag{9}$$

and as an approximation

$$T_0' - T_1' = \frac{1}{2}\frac{E}{c^2}v^2, \tag{10}$$

whereas, in view of Eqs. (7) and (8) he should have obtained

$$T_0' - T_1' = (m_0 - m_1)c^2\left[\frac{1}{(1 - v^2/c^2)^{1/2}} - 1\right], \tag{11}$$

which, in combination with Eq. (6) would have yielded

$$(E_0' - E_0) - (E_1' - E_1) = \frac{E}{(m_0 - m_1)c^2}(T_0' - T_1'), \tag{12}$$

or, considered as a difference of two relations,

$$E_0' - E_0 = \frac{E}{(m_0 - m_1)c^2}(T_0' + C)$$

and

$$E_1' - E_1 = \frac{E}{(m_0 - m_1)c^2}(T_1' + C). \qquad (13)$$

Comparing Eqs. (13) with Einstein's assumption, namely,

$$E_0' - E_0 = T_0' + C \quad \text{and} \quad E_1' - E_1 = T_1' + C,$$

(in Einstein's paper the kinetic energy T is denoted by K and the energy of radiation E by L), we see that Einstein unwittingly assumed that

$$\frac{E}{(m_0 - m_1)c^2} = 1, \qquad (14)$$

which is exactly the contention to be proved.[24]

The relativistic mass-energy equivalence in the case of a change of kinetic energy can easily be demonstrated as follows. The increment in kinetic energy dE_k is given by the product of force F and displacement ds:

$$dE_k = F\,ds = \frac{d(mv)}{dt}\,ds = v\,d(mv) = v^2\,dm + mv\,dv. \qquad (15)$$

Using the mass-transformation equation $m = m_0(1 - v^2/c^2)^{-1/2}$ we obtain, after squaring and rearranging,

$$m^2c^2 = m^2v^2 + m_0^2c^2 \qquad (16)$$

which after differentiation and cancellation of the common factor $2m$ gives

$$c^2\,dm = v^2\,dm + mv\,dv. \qquad (17)$$

Comparison with Eq. (15) shows that

[24] It should, however, be noted that Einstein proved the equivalence of mass and energy also in the following articles, in addition to the paper quoted in reference 19: "Prinzip von der Erhaltung der Schwerpunksbewegung und die Trägheit der Energie," *Annalen der Physik 20,* 627–633 (1906); "Die vom Relativitätsprinzip geforderte Trägheit der Energie," *ibid., 23,* 371–384 (1907); "Elementary derivation of the equivalence of mass and energy," *Bulletin of the American Mathematical Society 41,* 223–230 (1935).

$$dE_k = c^2\,dm, \tag{18}$$

and by integration

$$E_k = mc^2 - m_0c^2. \tag{19}$$

Thus, if m_0c^2 could be interpreted as some form of energy, as "energy of constitution" or "rest energy," or if all energy could be reduced to the kinetic form, mass and energy were only different names for the same physical existent. Although physics as yet could show only that electromagnetic energy of radiation and kinetic energy possess inertia, philosophers were eager immediately to generalize these results and to contend a universal identity of mass and energy. Among the most ardent proponents of these ideas were the advocates of the monistic school of thought who seized this opportunity in their search for a unified conceptual model of the universe. Thus, Gustave Le Bon, in his book *L'évolution de la matière,* published in 1905, spoke of a "dematerialization of matter into energy,"[25] and Julius von Olivier declared in 1906: "The mass of a body is synonymous with its energy."[26] Later, Bertrand Russell, in his well-known work *Human knowledge: its scopes and limits,* stated:

> Mass is only a form of energy, and there is no reason why matter should not be dissolved into other forms of energy. It is energy, not matter, that is fundamental in physics.[27]

But as so often happens, philosophy, in its inherent tendency toward generalization, anticipates and forestalls results, the corroboration and confirmation of which is still a matter of no little

[25] Gustave Le Bon, *L'évolution de la matière* (Flammarion, Paris, 1905). Hennemann's contention that Schelling's (1775–1854) "dynamic atomism" has "found in the equation $E = mc^2$ an amazing verification" is, in our view, an illegitimate retroprojection of modern scientific conceptions into the ideas of post-Kantian idealism which, by its speculative nature and obscure language, lends itself easily to many interpretations. See Gerhard Hennemann, *Naturphilosophie im 19. Jahrhundert* (Verlag Karl Alber, Freibourg and Munich, 1959), p. 36.

[26] Julius von Olivier, *Monistische Weltanschauung* (Naumann, Leipzig, 1906), p. 2.

[27] Bertrand Russell, *Human knowledge: its scopes and limits* (Simon and Schuster, New York, 1948), p. 291.

effort on the part of the exact sciences. As Whitehead once remarked, "philosophy builds cathedrals before the workmen have moved a stone."

In fact, it was still a long way until physics finally showed the universal identity of mass and energy. In 1907 Max Planck[28] gave a correct derivation of the mass-energy relation, based on Poincaré's momentum of radiation. In 1908 Comstock showed on the basis of electrodynamic considerations that $E = 3/4mc^2$ in conformity with Thomson's and Hasenöhrl's original results.[29] In the same year G. N. Lewis, in a paper entitled "A revision of the fundamental laws of matter and energy," [30] by inference from the theory of radiation pressure, derived the equation $dE = c^2\, dm$. Paul Langevin, in 1913, applied the mass-energy relation for the first time to nuclear physics in his explanation of the deviation of atomic weights from integral numbers.[31] Lenz,[32] Sommerfeld,[33] and Smekal [34] were the first to understand the general importance of this relation for nuclear physics.

The question to what degree of accuracy the early quantitative measurements of nuclear reactions confirmed the mass-energy relation was discussed in detail by Bainbridge in 1933.[35]

In spite of the general acceptance of the mass-energy relation as a formula of universal validity, it was still an open question whether all mass, without residuum, is transformable into energy.

[28] *Sitzungsberichte der Preussischen Akademie der Wissenschaften, physik.-mathem. Klasse 13* (June 1907).

[29] Daniel F. Comstock, "The relation of mass to energy," *Philosophical Magazine 15,* 1–21 (1908).

[30] *Philosophical Magazine 16,* 705–717 (1908).

[31] Paul Langevin, "L'inertie de l'énergie et ses conséquences," *Journal de physique théorique et appliquée 3,* 553–591 (1913).

[32] W. Lenz, "Über ein invertiertes Bohrsches Modell," *Sitzungsberichte der mathem.-phys. Klasse der K. B. Akademie der Wissenschaften zu München* (1918), pp. 355–365.

[33] Arnold Sommerfeld, *Atombau und Spektrallinien* (Brunswick, 1919), p. 538.

[34] Adolf Smekal, "Über die Dimensionen der α-Partikel," *Sitzungsberichte der Akademie der Wissenschaften in Wien 129,* 455–481 (1920).

[35] Kenneth T. Bainbridge, "The equivalence of mass and energy," *Physical Review 44,* 123 (1933).

The final answer, in the affirmative, was given in 1933 by Blackett and Occhialini[36] in their famous experiments on pair production and by its mirror phenomenon, the annihilation of matter. In 1934 Klemperer[37] demonstrated unequivocally that a positron and an electron can annihilate each other and produce energy of amount $2m_0c^2$ ($= 10^6$ ev). More recently, numerous experiments with antiprotons and antineutrons added further support to this contention. It is no exaggeration to say that the development of modern nuclear physics would have been impossible without the assumption of the mass-energy equivalence. Braunbeck[38] showed in 1937 that the experimental verification of this relation is so well established that the mass-energy equivalence should no longer be regarded as a theorem, derivable from other principles of less direct and less accurate empirical evidence, but should be taken, like the energy- conservation law, as one of the fundamental principles in physics.

The new insight gained from the theory of relativity and corroborated by nuclear physics threw new light on the concept of energy. In the first place, since matter was transformable into energy, energy lost its classical indeterminateness as far as an additive constant was concerned. It became a physical quantity of absolute magnitude. Secondly, the electromagnetic concept of mass appeared in a new light. Now it was possible to understand what Thomson, Heaviside, Lorentz, Wien, and Abraham actually had been doing. Now it became clear that they had been playing with relativistic field dynamics.[39] By the displacement of an electric charge the electrostatic field gives rise to a magnetic

[36] P. M. S. Blackett and G. P. S. Occhialini, *Proceedings of the Royal Society, London 139*, 699 (1933).

[37] O. Klemperer, "Annihilation radiation of the positron," *Proceedings of the Cambridge Philosophical Society 30*, 347–354 (1934); M. Deutsch, "Annihilation of swift positrons," *Physical Review 72*, 729–730 (1947).

[38] Werner Braunbeck, "Die empirische Genauigkeit des Mass-Energie Verhältnisses," *Zeitschrift für Physik 107*, 1–11 (1937).

[39] It will be recalled that Maxwell's equations are Lorentz covariant if the quantities involved are regarded as components of world vectors or world tensors.

field and both fields produce a flux of energy and of momentum. It is, therefore, not surprising that the proponents of the electromagnetic concept of matter inferred the existence of an inertia proportional to E/c^2. The problem, of course, still to be solved was this: why did they conclude that $m = (4/3)E/c^2$ [40] and not, as relativity demands, $m = E/c^2$?

It is interesting to note that more than 15 years had passed before some progress was achieved in clarifying this discrepancy — and perhaps even today the issue is not yet finally resolved in all its aspects. In 1922 Fermi, in a paper entitled "Correction of a serious discrepancy between the theory of electromagnetic mass and the theory of relativity," [41] and later, in 1936, Wilson, in an article "The mass of a convected field and Einstein's mass-energy law," [42] tried to solve the problem by taking into account stresses of the material support of the field and elastic tensions in the surrounding medium. Just as the energy flow in a stressed belt (as mentioned above in connection with Mie's theory) constitutes a flux in a direction opposite to that of the motion, so does the motion of a charged particle produce an energy current that contributes to the total momentum a negative term. The inertia of the system is thereby reduced by a (negative) correction term of approximately the amount $e^2/6ac^2$, that is, one-third of the field energy of the particle, so that the value of the total inertia, in conformity with relativity, turns out to be $m = E/c^2$. This approach of reconciling the Abraham-Lorentz electromagnetic theory of mass with the theory of relativity remains, however, unsatisfactory insofar as the separation of nonelectromagnetic from electromagnetic forces is not Lorentz-invariant. The ultimate reason for this incompatibility of the Abraham-Lorentz theory with our modern point of view lies in Abraham's

[40] See, for example, Heaviside's result, Chapter 11, Eq. (6).

[41] Enrico Fermi, "Correzione di una grave discrepanza tra la teoria delle masse elettromagnetiche e la teoria della relativitá," *Rendiconti dei Lincei 31,* 184–187, 306–309 (1922).

[42] W. Wilson, "The mass of a convected field and Einstein's mass-energy law," *Proceedings of the Physical Society 48,* 736–740 (1936).

definition of the electromagnetic mass in terms of the field energy integrated over the space only outside the (variable) volume of the (moving) particle, whereas the relativistically correct definition ought to be based on the total energy-momentum tensor, which includes the matter tensor.

As we have seen, the conversion of an electron-positron pair, for example, into gamma radiation or its mirror phenomenon is an incontestable experimental confirmation of the assertion of the theory of relativity that mass and energy are mutually and completely interconvertible.

This state of affairs raises the following fundamental questions: Are not the two entities which are interchangeable essentially the same? Is not what is generally spoken of as an equivalence in reality an identity? Are therefore not "mass" and "energy" merely synonyms for the same physical reality, which — in analogy to the term "wavicle" coined by Eddington to describe the electron in view of the wave-particle parallelism in wave mechanics — may perhaps be termed "massergy"?

In order to give a satisfactory and comprehensive answer to these crucial questions the following considerations have to be taken into account. In prerelativistic physics three basic conservation theorems were of predominant importance: (1) conservation of momentum (Newton's law of conservation of the center of gravity; see Chapter 12), (2) conservation of mass (Lavoisier's law; see Chapter 8), and (3) conservation of energy (Robert Mayer, 1842, and Hermann Helmholtz, 1847). Momentum being a three-dimensional vector quantity, these conservation laws constitute five equations or conditions that every physical process has to comply with.

In the theory of relativity there is only one conservation law instead: the momentum-energy four-vector P^{μ} is conserved. These are four equations or conditions. Thus, instead of the two separate prerelativistic conservation laws of mass *and* energy there is in the theory of relativity only *one* conservation law of mass *or* energy ("massergy"). In particular, the rigorous validity of

Lavoisier's law is renounced. Since the heat of reaction Q as energy also has mass, it is clear that in an exothermic reaction the (prerelativistic) "mass" decreases and in an endothermic reaction it increases.

Consider, for example, a kinetic reaction between two interacting molecules of masses m_{01} and m_{02}, emerging after the reaction with masses m_{03} and m_{04}. For simplicity we assume all motions to take place in the direction of the x-axis so that the momentum-conservation theorem contains only one equation. Let us describe the phenomenon by a prerelativistic representation and by a relativistic representation.

In the prerelativistic representation, we have:

Conservation of mass or Lavoisier's law:

$$m_{01} + m_{02} = m_{03} + m_{04}, \tag{20}$$

Conservation of momentum:

$$m_{01}v_1 + m_{02}v_2 = m_{03}v_3 + m_{04}v_4, \tag{21}$$

Conservation of energy:

$$\tfrac{1}{2}m_{01}v_1^2 + \tfrac{1}{2}m_{02}v_2^2 = \tfrac{1}{2}m_{03}v_3^2 + \tfrac{1}{2}m_{04}v_4^2 + Q. \tag{22}$$

In these equations, v_n is the velocity of the particle having the mass m_{0n}, and Q is the (positive or negative) heat of reaction. If m_{0n} and v_n is given for $n = 1, 2, 3$, then m_{04} is determined by Eq. (20), v_4 by Eq. (21), and Q by Eq. (22).

The relativistic representation is:

Conservation of the energy-momentum four-vector:

$$P_1^\mu + P_2^\mu = P_3^\mu + P_4^\mu. \tag{23}$$

If U_n^k is the k-component of the four-velocity of the particle having the proper mass m_{0n}, Eq. (23) gives for the spatial components, $\mu = 1$:

$$m_{01}U_1^1 + m_{02}U_2^1 = m_{03}U_3^1 + m_{04}U_4^1. \tag{24}$$

Since in our case $U_n^k = v_n(1 - v_n^2/c^2)^{-1/2}$, we obtain

$$\frac{m_{01}v_1}{(1 - v_1^2/c^2)^{1/2}} + \frac{m_{02}v_2}{(1 - v_2^2/c^2)^{1/2}} = \frac{m_{03}v_3}{(1 - v_3^2/c^2)^{1/2}} + \frac{m_{04}v_4}{(1 - v_4^2/c^2)^{1/2}}, \quad (25)$$

and for $\mu = 4$ (the temporal components)

$$\frac{m_{01}c^2}{(1 - v_1^2/c^2)^{1/2}} + \frac{m_{02}c^2}{(1 - v_2^2/c^2)^{1/2}} = \frac{m_{03}c^2}{(1 - v_3^2/c^2)^{1/2}} + \frac{m_{04}c^2}{(1 - v_4^2/c^2)^{1/2}}. \quad (26)$$

Equation (25) expresses the conservation of momentum and Eq. (26) the conservation of mass or energy ("massergy").[43] It should be noted that in Eq. (26) no additional term for Q can be posited, since the equation states the conservation of the total energy. Let us now assume for a moment that Lavoisier's law

$$m_{01} + m_{02} = m_{03} + m_{04} \quad (27)$$

holds also in relativistic physics. Again, let m_0 and v_n be given for $n = 1$, 2, and 3. Then m_{04} is determined by Eq. (20), and v^4 by Eq. (25). But Eq. (26) would then in general be inconsistent with the preceding equations. Thus, relativistic dynamics, recognizing the variability of proper mass in interaction processes among particles of a system, disproves Lavoisier's law.

It is interesting to note that at the end of the last century, years before the advent of relativity, the rigorous validity of Lavoisier's law was already being seriously questioned. D. Kreichgauer, as the result of certain experiments on reactions involving mercury, bromine, and iodine,[44] was perhaps the first, in 1890–1891, to express doubts concerning the absolute correctness of Lavoisier's law. Not much later H. Landolt began his series of experiments on the reduction of silver sulfate, in the course of which he thought he found perceptible differences

[43] In the derivation of Eq. (26) from Eq. (24) use has been made of the following relations:

$$P_1^4 = m_{01}U_1^4 = m_{01}\frac{dx_1^4}{ds} = \frac{m_{01}}{(1 - v_1^2/c^2)^{1/2}} \frac{d(ict)}{c\,dt} \propto \frac{m_{01}c^2}{(1 - v_1^2/c^2)^{1/2}}$$

and corresponding relations for P_2^4, P_3^4, and P_4^4.

[44] *Verhandlungen der physikalischen Gesellschaft zu Berlin 10*, 13–16 (1891).

between the masses before and those after the reaction.[45] These differences, he insisted, were greater than the error that could possibly be ascribed to his methods of weighing. The problem was further investigated by M. Haensel[46] and by A. Heydweiller,[47] who came to the conclusion that "it can be stated with certainty that differences in mass occur in the reaction of iron on copper sulfate."[48] Others who believed firmly in the validity of Lavoisier's law tried to explain these discrepancies as secondary effects. Thus, for instance, it was claimed that those among these experiments which involved iron compounds are manifestations of a still unknown connection between electromagnetic and gravitational forces, a relation that Faraday had already considered possible.[49] Later, however, it was realized that buoyancy effects, originating from the thermal expansion of the enclosures of the specimens due to the heat of reaction, had made the weighing procedures inaccurate.[50]

Today we know undoubtedly that no ordinary weighing procedures were, or even at present are, sensitive enough to show the weight or mass corresponding to the heat of reaction Q in a chemical reaction.[51] In fact, it was instrumental errors and errors in method that were responsible for the suspicions entertained as to Lavoisier's law at the turn of the century. It must, how-

[45] H. Landolt, "Untersuchungen über etwaige Änderungen des Gesamtgewichtes chemisch sich umsetzender Körper," *Zeitschrift für physikalische Chemie 12*, 1–11 (1893). See also *Naturwissenschaftliche Rundschau 15*, 66 (1900).

[46] *Inaugural-dissertation* (Breslau, 1899).

[47] Adolf Heydweiller, "Über Gewichtsänderung bei chemischer und physikalischer Umsetzung," *Annalen der Physik 5*, 394–420 (1901).

[48] "Als sicher festgestellt kann man also die Gewichtsänderung betrachten: bei der Wirkung von Eisen auf Kupfersulfat in saurer oder basischer Lösung, bei der Auflösung von saurem Kupfersulfat und bei der Wirkung von Kaliumhydroxyd auf Kupfersulfat." *Ibid.*, p. 417.

[49] F. Sanford and L. Ray, *Physical Review 5*, 247 (1897); *7*, 236 (1898).

[50] See D. Pekar, "Das Gesetz der Proportionalität von Trägheit und Gravitation," *Naturwissenschaften* (7. Jahrgang, 1919), p. 327; see also W. Roth, *ibid.*, p. 416.

[51] In the exothermic reaction $2\,H_2 + O_2 \rightarrow 2\,H_2O + Q$, the mass of Q (117 cal) is of the order of 10^{-12} gm for a reacting mass of 36 gm.

ever, be admitted that as a result of those investigations the a priori necessity of Lavoisier's law was early called in question. Thus, for example, Wilhelm Ostwald, in one of his lectures on natural philosophy, held at the University of Leipzig in 1901, declared with reference to Lavoisier's law:

> Whether this law holds rigorously, cannot be said a priori. The various attempts to present the law as a logical necessity — as such a profound thinker as Schopenhauer among others did — are reminiscent of the ontological proof. For the train of thought is as follows: matter is conceived as the unchangeable carrier of variable qualities; therefore mass, as the fundamental property of matter, must itself be unchangeable. In the same manner one could prove the paradoxical statement that matter cannot change its place, since occupying space is likewise a fundamental property of matter.[52]

After this historical digression let us return to our principal question. In view of the fact that in relativity there is only one conservation law of mass or energy ("massergy"), the rigorous answer to these questions undoubtedly is: mass and energy are identical, they are synonyms for the same physical substratum.

However, nature does not always manifest itself in all the rigor of its theoretical possibilities. In the vast majority of known physical processes — apart from nuclear phenomena and interaction processes of fundamental particles — the mass of a body, as manifested by weighing procedures, and its energy, as determinable by the work done or the heat produced or absorbed, are conserved separately. Nature ordinarily behaves as if two conservation laws exist.

There were times when a wealthy businessman had two separate bank accounts: a long-term account comprising most of his wealth and not affected by his daily commercial transactions, and a short-term account undergoing rapid fluctuations from day to day. Similarly, "massergy" in the overwhelming majority of physical processes under ordinary conditions, has two layers

[52] Wilhelm Ostwald, *Vorlesungen über Naturphilosophie* (Veit, Leipzig, ed. 3, 1905), p. 283.

or modes of manifestation: a passive or bound mode that does not take part in what is generally called "energy transformation," and an active mode that undergoes these transformations.

Since the predominant part of the "massergy" of a body belongs to the passive component and only a negligibly small fraction to the active one, the insufficient sensitivity of weighing procedures simulates a conservation of mass. Correspondingly, calorimetric methods or similar procedures simulate, in a complementary manner, a conservation of energy. Thus, from the practical point of view, a retention of the prerelativistic division into mass and energy seems to some extent still justified.

One has, however, to keep in mind that the partition of "massergy" into a passive component (mass) and an active component (energy) depends on the circumstances under discussion. In ordinary chemical reactions the energy of the inner electronic shells — in addition to the intranuclear energy — is part of the passive component, while in high-temperature processes (under complete atomic ionization) it is part of the active component. In nuclear reactions, involving rearrangements of nucleons, the intranuclear energy becomes part of the active component while the proper masses of the nucleons still form the stock of the passive component. Finally, in the physics of elementary particles the division into mass and energy loses its meaning since such particles can transmute their proper masses into energy either *in toto* or not at all.

The prerelativistic differentiation between mass and energy as two mutually exclusive and fundamentally distinct categories of existence is rooted — like numerous other concepts — also in the physiological-biological nature of the human body. The sensitivity of the human body with respect to energy perceptions, such as electromagnetic radiation for vision or acoustical vibrations for audition, is much higher than its sensitivity with respect to mass perceptions, that is, its response to the inertial or gravitational effects of masses. Consider, for example, the sensitivity of the retina of the human eye to radiant energy. It has been shown[53] that rod vision at the wavelength of peak sensitivity (about 5100 A) can be called into being by about 5 quanta, corresponding to about 10^{-11} erg, whereas tactile perception of the inertial or gravitational effects of, say, 1 gram amounts to about 10^{21} erg (1 gram $= c^2$ erg). The human sensory apparatus is thus 10^{32} times more sensitive to energy perceptions than it is to mass perceptions.

If this ratio were of the order of unity, and not 10^{32}, the identity of mass and energy would have been an obvious fact of experience. The human eye, perceiving light from the sun, would then also feel the impact of the photons. Contiguity in time and space would thus induce the human mind to look upon mass and energy as only different aspects of one and the same physical existent.[54]

In conclusion of this chapter the question may be asked: Why is it that in ordinary physico-chemical processes the vast majority of the "massergy" of a body remains latent and inactive and only an infinitesimally small amount takes part in "transformations of energy"? Why does not the physical substratum always behave as it does in the physics of elementary particles? The answer to this question lies beyond the scope of the theory of relativity. It lies in the quantum-mechanical aspects of the dynamics of the atom, in the existence of stationary states and discrete energy levels. This is the ultimate reason for the retention and survival of the classical notion of mass. It is a *phenomenon bene fundatum,* as Leibniz once characterized mass.

[53] S. Hecht, S. Shlär, and M. H. Pirenne, "Energy, quanta, and vision," *Journal of General Physiology 25,* 819 (1942).

[54] See Friedrich Dessauer, "Kleine Notiz über den menschlichen Standort gegenüber dem Materie-Energieproblem," *Helvetica Physica Acta 15,* 108–110 (1942).

CHAPTER **14**

THE CONCEPT OF MASS IN QUANTUM MECHANICS AND IN FIELD THEORY

Although the logico-mathematical foundations of non-relativistic quantum mechanics of a finite number of degrees of freedom are, from the technical point of view at least, sufficiently established for the consistent formulation of a deductive theory, the semantic interpretation and, in particular, the methodological role of observables are still matters of some ambiguity. Specifically, the logico-methodological status of the concept of mass within the structure of classical quantum mechanics seems never to have been thoroughly clarified.

The usual procedure is to carry the notion of mass over from classical mechanics right into the various representations of quantum mechanics. Mass thus appears as an extraneous parameter in the formulation of quantum-mechanical problems which are treated by means of operators and state functions. The question whether mass itself is not an observable or something immediately derivable from observables and whether it consequently has not to be represented by an operator like all other observables, is generally blithely overlooked.

It may be argued that the definition and determination of mass after the Newtonian manner may legitimately be adopted for quantum mechanics, since the Newtonian equations of motion, according to certain well-known theorems by Ehrenfest and others, or according to the correspondence principle, are, at least as "average theorems," a logical consequence of the fundamental equations of quantum dynamics.[1] It must, however, be realized that Ehrenfest's theorems reveal only an analogy between the dynamical behavior of wave packets and that of Newtonian particles; they do not reduce conceptually the former to the latter. The correspondence principle, on the other hand, in spite of its historic and heuristic importance for the development of quantum theory, is not an integral part of the theory itself. Indeed, a rigorous derivation from classical physics of the fundamentals of quantum mechanics is impossible — in spite of the fact that Schrödinger's wave mechanics bears to classical particle mechanics the same relation as physical optics to geometric optics. A logical derivation of Schrödinger's equation from classical mechanics does not, and cannot, exist.

It would perhaps be less objectionable to regard m in the fundamental equations of quantum mechanics at the beginning merely as a parameter characteristic of the special type of particles to which the equation refers, and to deduce its physical significance as an inertial factor in the subsequent development of the theory. The preliminary interpretation of the parameter m and its operational determination could be given by de Broglie's equation for a diffraction process,

$$\lambda = \frac{h}{mv}, \tag{1}$$

where the wavelength λ is measured from the diffraction pattern and v by one of the conventional methods of velocity determination. The parameter m in Eq. (1) has for the time being no inertial significance. The question then arises whether the interpretation of m as an inertial factor can be derived from the wave properties

[1] P. Ehrenfest, "Bemerkung über die angenäherte Gültigkeit der klassischen Mechanik innerhalb der Quantenmechanik," *Zeitschrift für Physik 45*, 455–457 (1927). See also A. E. Ruark, *Journal of the Optical Society of America 16*, 40–43 (1928) and *Physical Review 31*, 533–538 (1928).

of matter and radiation alone. In fact, it can be shown[2] that de Broglie's equation, the energy-frequency relation $E = h\nu$, and the Lorentz transformation equations lead to an inertial interpretation of m in Eq. (1). Interesting as such a derivation of inertial properties of matter from its wave aspects seems to be, it is not a satisfactory procedure for introducing the notion of mass into classical quantum mechanics, because inertial mass here appears as a relativistic effect.

An approach that is more consistent and congenial with the spirit of classical quantum mechanics is perhaps to assign a "density" directly to the wave function with a distribution over the nominally infinite wave front. Integrating this "density" over all three-dimensional space would then yield the mass value of the particle represented by the wave or wave packet. Such a definition of "mass" in quantum mechanics was, in fact, employed by Eddington[3] in his reformulation of the exclusion principle by assigning a saturation value for the "density" of an elementary wave function.

The corrections advanced by Fermi and Wilson in their attempt to reconcile the electromagnetic interpretation of mass with the theory of relativity included, as we have seen, certain considerations extraneous to electromagnetic theory. It may be asked whether such a recourse to extraelectromagnetic considerations has to be interpreted as an indication that the electromagnetic theory alone is unable to account for the inertial properties of matter.

Let us investigate this problem from the viewpoint of modern field theory. In classical physics the mass of a particle, on the whole, was a notion that was independent of the notion of the field. Particles and fields were regarded as two essentially dif-

[2] Richard Schlegel, "Wave and inertial properties of matter," *American Journal of Physics 22*, 77–82 (1954).

[3] Sir Arthur Eddington, "A new derivation of the quadratic equation for the masses of the proton and electron," *Proceedings of the Royal Society 174*, 16–41 (1940).

ferent agencies. Particles were the sources of the field and were acted on by the field but were not part of the field. Mass appeared in the equations of motion in the form of a parameter that was characteristic for the particle under discussion and that described its inertial behavior. Mass, in this sense, will be called "the mechanical mass" of the particle and denoted by m_m. The electromagnetic mass discussed in Chapter 11 was, in fact, an exception since its origin was supposed to lie exclusively in the interaction between the charge of the particle and the electromagnetic field. Mass, in this theory, was a derived notion, but charges and fields were still mutually irreducible and essentially different agencies. The mass of a particle that is supposed to originate in the interaction with the field of the particle (the "self-field"), or is assumed to be produced by the field alone, will be called the "field mass" and will be denoted by m_f.

A theory in which the total mass of a particle is its field mass will be called "mass-unitary." Following Born and Infeld[4] we shall call a theory "unitarian" if it postulates the existence of only one physical entity. A unitarian field theory is always mass-unitary, but not every mass-unitary theory is necessarily unitarian. The electromagnetic theory of mass, for example, was an attempt to formulate a mass-unitary theory, but it was not a unitarian theory since charges and fields were mutually irreducible concepts.[5] A mass-unitary field theory has to satisfy the following conditions:

(1) The total energy U_f of the field produced by the particle must be finite (that is, a nondivergent quantity).

(2) U_f/c^2 must be equal to the experimentally determined value of the mass of the particle.

[4] M. Born and L. Infeld, "Foundations of the new field theory," *Proceedings of the Royal Society 144*, 425 (1934).

[5] The terminology in scientific literature in this respect is not consistent. French theorists usually employ the term "théorie unitaire" in the sense of the English "unified theory"; see, for instance, A. Lichnerowicz, *Théories relativistes de la gravitation et de l'électromagnétisme* (Paris, 1955), p. 150. J. Rzewuski, *Field theory* (Warsaw, 1958), p. 259, uses "unitarian" in the sense of our "mass-unitary" and so forth.

(3) The momentum G_f of the field must be finite (that is, a nondivergent quantity).

(4) G_f and U_f must form a four-vector.

(5) The theory must lead to the equations of motion of the particle.

(6) The theory must lead to the spin value of the particle as determined by experiment.

Since the spin of a particle is a quantum-mechanical effect, condition (6) applies only to quantum-field theories and not to classical field theories which we shall discuss first.

In order to examine whether the classical theory of the electromagnetic field is a mass-unitary theory we recall that the momentum $\mathbf{P}_m$ and the energy E (or more precisely, $(i/c)E$) of the particle (the electron) form a four-vector. If F_{ik} is the electromagnetic field tensor[6] and f_k the Lorentz force density, so that

$$f_k = \frac{1}{4\pi} F_{kn} \frac{\partial F_{nr}}{\partial x_r}, \tag{2}$$

it will be recalled that f_k is related to the energy-momentum tensor of the electromagnetic field T_{mn} by the equation

$$f_k = \frac{\partial T_{kn}}{\partial x_n}, \tag{3}$$

where

$$T_{kn} = \frac{1}{4\pi} (F_{kr}F_{rn} + \tfrac{1}{4}\delta_{kn}F_{qp}F_{qp}). \tag{4}$$

Integrating the left-hand member of Eq. (3) over the invariant four-volume $d\Omega = dx_1\, dx_2\, dx_3\, dx_4$ ($dx_4 = icdt$), we obtain

$$ic \iiiint f_1\, dV\, dt = ic \int F_x\, dt = icG_{mx}, \tag{5}$$

where F_x is the x-component of the force exerted on the particle and $G_{mx} = P_{mx}$ is the x-component of the (mechanical) momentum of the particle. For the time component we obtain

[6] See L. Landau and E. Lifshitz, *The classical theory of fields* (Addison-Wesley, Reading, Mass., 1951), pp. 56–88. Repetition of indices implies summation.

$$ic \iiiint f_4 \, dV \, dt = -\int w \, dV = -E, \tag{6}$$

where w denotes the energy density. Since the integration of a four-vector over an invariant volume $d\Omega$ does not destroy the tensor character of the four-vector, it is clear again that the momentum and the energy of the particle combine into a four-vector.

Calculating now the integral of the right-hand member of Eq. (3), we obtain for the space components

$$\begin{aligned} ic \iiiint \frac{\partial T_{1n}}{\partial x_n} \, dV \, dt &= ic \iiiint \left(\frac{\partial T_{1x}}{\partial x} + \frac{\partial T_{1y}}{\partial y} + \frac{\partial T_{1z}}{\partial z} + \frac{\partial T_{14}}{ic \, dt} \right) dV \, dt \\ &= ic \int dt \oint T_{1N} \, d\sigma - \frac{i}{c} \int S_x \, dV \\ &= ic \int dt \oint T_{1N} \, d\sigma - icG_x^{(f)}, \end{aligned} \tag{7}$$

where the integral over the divergence has been transformed into a surface integral with the help of Gauss's formula and where S_x is the x-component of the Poynting vector and $G_x{}^{(f)}$ the x-component of the field momentum. For the time component we obtain by a similar computation the expression

$$\int dt \oint S_N \, d\sigma + E^{(f)}, \tag{8}$$

where $E^{(f)} = U_f$ is the total energy of the field. Equating (7) and (8) with (5) and (6) we see that

$$P_{mx} = G_{mx} = -G^{(f)} + \int dt \oint T_{xN} \, d\sigma, \tag{9}$$

$$-E = E^{(f)} + \int dt \oint S_N \, d\sigma. \tag{10}$$

The integrals in the last two equations represent time integrations over the energy and momentum fluxes through the surface and thus depend on the past history of the system at all previous times. They cannot be interpreted as momentum or energy of a field. Consequently, $G^{(f)}$ and U_f, the momentum and the energy of the electromagnetic field, are themselves not components of a four-vector and thus cannot be identified with the momentum and the energy of the particle. In other words, the field mass m_f cannot be identified with the mechanical mass m_m and the classical theory of the electromagnetic field is not mass-unitary.

The result that in the case of the electromagnetic field the field momentum $G_1^{(f)} = G_x^{(f)}, \ldots, G_4^{(f)} = E^{(f)} = U_f$ is not a four-vector follows also immediately from a theorem which according to Gustav Mie[7] is usually referred to as "Laue's theorem". If $T_{kn}{}^0$ is the energy-momentum tensor and dV^0 the three-dimensional volume with respect to a reference system in which the charged particle (electron) that produces the field is at rest, then $G_k^{(f)}$ is a four-vector if and only if $\int T_{rs}{}^0 dV^0 = 0$ for all $r = s$ with the exception of $r = s = 4$.

Now, in a system relative to which the particle is at rest, the magnetic field vector vanishes and the total field energy density is $T_{44}{}^0 = (1/8\pi)E^2$ and, as we know from the Maxwell stress tensor, $T_{11}{}^0 = (1/4\pi)(E_x^2 - \frac{1}{2}E^2)$. Thus, $\int T_{44}{}^0 \, dV^0 = (1/8\pi) \int E^2 \, dV = U^0$ and $\int T_{11}{}^0 \, dV^0 = -\frac{1}{3}U^0 \neq 0$. Consequently, Laue's theorem does not apply for the electromagnetic field, $G_k^{(f)}$ is not a four-vector, and condition (4) for mass-unitary fields is not satisfied.

Any attempt to arrive at a mass-unitary field theory must therefore start with a thoroughgoing modification of Maxwell's fundamental equations of the electromagnetic field. Most of these reformulations of the Maxwell-Lorentz theory, it must be admitted, were not concerned so much with a consistent mass-unitary derivation of the electronic mass but rather were carried

[7] Gustav Mie, "Grundlagen einer Theorie der Materie," *Annalen der Physik* *40*, 7 (1913). Laue's theorem can be proved by means of the transformation equations of the energy-momentum tensor: $T_{14} = i\beta\gamma^2(T_{11}{}^0 - T_{44}{}^0)$ and $T_{44} = \gamma^2(T_{44}{}^0 - \beta^2 T_{11}{}^0)$, where in view of the spherical symmetry of the rest field only diagonal terms are to be taken into account. Here $G_k^{(f)}$ is defined by $(i/c)\int T_{k4} \, dV$. Thus $G_k{}^0 = 0$ for $k = 1, 2, 3$ and $G_4{}^0 = (i/c)\int T_{44} \, dV = (i/c)E^{(f)}$ If $G_k^{(f)}$ is a four-vector it satisfies the transformation equations $G_1^{(f)} = -i\beta\gamma G_4{}^0 = (\beta\gamma/c)\int T_{44}{}^0 \, dV^0$ and $G_4^{(f)} = (i\gamma/c)\int T_{44}{}^0 \, dV^0$. On the other hand, we have $G_1^{(f)} = (i/c)\int T_{14} \, dV = (\gamma\beta/c)\int (T_{44}{}^0 - T_{11}{}^0) \, dV^0$ and $G_4^{(f)} = (i\gamma/c)\int (T_{44}{}^0 - \beta^2 T_{11}{}^0) \, dV^0$ with $dV = \gamma \, dV^0$. Comparison shows that $\int T_{11}{}^0 \, dV^0 = 0$.

out with a slightly different objective. They were motivated by the desire to overcome certain difficulties of a related nature in quantum electrodynamics. Modern quantum electrodynamics as developed by Dirac,[8] Heisenberg and Pauli,[9] Jordan, and Fermi, based on the conceptions of an electromagnetic field and an electron-positron field, is both a field theory and a theory of elementary particles. States of the quantized electromagnetic field are associated with photons and states of the quantized electron-positron field are associated with electrons. At an early stage of the development of this theory, however, it was realized that the self-energy of the electron, at the lowest order of the perturbation-theoretic computation, turns out to be a divergent quantity.[10] And although this divergence, on the basis of Dirac's hole theory, owing to the suppression of the part of the self-energy of the vacuum, was only of a logarithmic nature — in contrast to the linear divergence of the classical self-energy — the unavoidable consequence of an infinite mass for the elementary particle seemed to make the theory useless for all practical purposes. In order to facilitate the solution of these divergence difficulties, it was thought advisable first to revise classical electromagnetic theory so as to get rid of these infinities in the classical realm, and it was hoped that a quantization of the modified classical theory would then lead to a consistent quantum electrodynamics.

One of the most promising approaches at that time seemed to be Born's nonlinear theory of the electromagnetic field.[11] It

[8] P. A. M. Dirac, "The quantum theory of the emission and absorption of radiation," *Proceedings of the Royal Society 114,* 243–265 (1927); "The quantum theory of dispersion," *ibid.,* 710–728.

[9] W. Heisenberg and W. Pauli, "Zur Quantendynamik der Wellenfelder," *Zeitschrift für Physik 56,* 1–61 (1929).

[10] I. Waller, "Bemerkungen über die Rolle der Eigenenergie des Elektrons in der Quantentheorie der Strahlung," *Zeitschrift für Physik 62,* 673–676 (1930); J. R. Oppenheimer, "Note on the theory of the interaction of field and matter," *Physical Review 35,* 461–477 (1930).

[11] Max Born, "On the quantum theory of the electromagnetic field," *Proceedings of the Royal Society 143,* 410–437 (1934); Max Born and Leopold Infeld, "Foundations of the new field theory," *ibid. 144,* 425–451 (1934). The

will be recalled that the field equations are a consequence of the principle of least action, according to which the integral over the Lagrangian has a stationary value.[12] In view of the invariance of $E^2 - H^2$ Born postulated as the Lagrangian the function

$$L_0 = \frac{E_0^2}{4\pi}\left[1 - \left(1 - \frac{E^2 - H^2}{E_0^2}\right)^{1/2}\right], \tag{11}$$

where E_0 is the so-called "maximum field." For $E \ll E_0$ and $H \ll H_0$, L_0 approaches the classical Lagrangian $(1/8\pi)(E^2 - H^2)$ of the electromagnetic field. For a purely electrostatic field with an interaction between the electron and the field, Eq. (11) has to be modified to

$$L = \frac{E_0^2}{4\pi}\left[1 - \left(\frac{E^2}{E_0^2}\right)^{1/2}\right] - e\varphi\,\delta(r), \tag{12}$$

where φ is the interaction potential, e the charge, and $\delta(r)$ Dirac's function. From Eq. (12) the components of the energy-momentum tensor can be calculated, with the result that

$$\int T_{11}^0\,dV^0 = \frac{e^2}{r_0}\left(\frac{1}{3}c_1 - c_2\right), \tag{13}$$

$$\int T_{44}^0\,dV^0 = \frac{e^2}{r_0}(c_1 - c_2), \tag{14}$$

where

$$r_0 = \left(\frac{e}{E_0}\right)^{1/2}, \quad c_1 = \int_0^\infty \frac{x^2}{(1 + x^4)^{1/2}}\,dx,$$

and

$$c_2 = \int_0^\infty x^2\left[1 - \frac{x^2}{(1 + x^4)^{1/2}}\right]dx.$$

Partial integration of the expression for c_2 shows that $c_2 = \frac{1}{3}c_1$. Equation (13) now shows that according to Laue's theorem con-

idea of the possibility of nonlinear field theories can be traced back to Gustav Mie's paper: "Grundlagen einer Theorie der Materie," *Annalen der Physik 37,* 511–534 (1912); *39,* 1–40 (1912); *40,* 1–66 (1913).

[12] Cf., for instance, L. Landau and E. Lifshitz, *op. cit.,* pp. 66–88.

dition (4) for mass-unitary fields is satisfied. The field mass m_f can be computed from Eq. (14) as follows:

$$m_f = \frac{1}{c^2} \int T^0_{44} \, dV^0 = \frac{2}{3} \frac{e^2}{r_0} \frac{c_1}{c^2} \approx 1.24 \frac{e^2}{c^2 r_0}. \tag{15}$$

If for m_f the experimental value of the proper mass of the electron is substituted ($m_f = 9.1085 \times 10^{-28}$ gm), the constant r_0 turns out to be equal to the classical radius of the electron, $r_0 \approx 3.5 \times 10^{-13}$ cm.[13] Conditions (1), (2), and (3) for mass-unitary fields are therefore satisfied. Finally, condition (5) is also satisfied, since in a nonlinear theory the law of motion is a consequence of the field equations, as is known from the general theory of relativity.[14] It is therefore clear that Born's nonlinear theory of the electromagnetic field is a mass-unitary theory.

As to the question of an experimental confirmation of the theory, it should be noted that the nonlinear effects that originate from the absence of a superposition principle, such as the scattering (or reflection) of radiation by radiation, are of extremely small magnitude and either beyond the limit of experimental verification or empirically explicable.[15] But the main reason why Born's nonlinear theory was a disappointment was not the experimental issue but its inadequacy and unsuitability for a quantum-mechanical generalization which could remove the divergence infinities not only of the so-called "longitudinal" self-energy, associated with classical field theory, but also of the "transverse" self-energy due to the quantum-mechanical interaction of a charged particle with the vacuum fluctuations.

[13] In Born's theory the electron is a point particle.

[14] See A. Einstein, L. Infeld, and B. Hoffmann, "The gravitational equations and the problem of motion," *Annals of Mathematics 39*, 65 (1938). See also Max Jammer, *Concepts of force* (Harvard University Press, Cambridge, 1957), pp. 262–263.

[15] Thus, for example, the collision of two photons or quanta of radiation can give rise to a virtual pair production and a subsequent annihilation to the restoration of the two photons with an over-all effect equivalent to the scattering of radiation by radiation.

Another approach to overcoming the divergence difficulties of m_f for a point particle which also originated in a classical procedure of eliminating singularities due to the electromagnetic self-energy of a point source, without destroying the validity of Maxwell's equations in the vicinity of the source and without impairing the relativistic invariance, was the "λ-limiting process," introduced in 1933 by Gregor Wentzel.[16] By the use of separate time coordinates for each charged particle and for the field and by the use of the Dirac-Fock-Podolsky formalism, a form factor is chosen in such a way that the integrated energy of the field is equal to zero. The process thus amounts to an elimination of m_f altogether. Other suggestions with the same objective in mind introduced additional compensating fields so as to nullify the total self-energy.[17]

Most revolutionary of the suggestions for overcoming the divergence difficulties of mass and energy were probably those that called for a fundamental revision of the applications of the concepts of space and time with respect to the fundamental particles. As early as 1930 Ambarzumian and Iwanenko[18] questioned the concept of a spatial expansion of elementary particles, introduced the idea of a cubic space lattice with a finite lattice constant, and outlined a program of replacing differential equations in physical theory by difference equations. Watagin[19] introduced the idea of an elementary length under the guise of

[16] Gregor Wentzel, "Über die Eigenkräfte der Elementarteilchen," *Zeitschrift für Physik 86,* 479–494 (1933); *87,* 726–733 (1934); "Recent research in meson theory," *Reviews of Modern Physics 19,* 1–18 (1947).

[17] E. C. G. Stückelberg, "Un nouveau modèle de l'électron ponctuel en théorie classique," *Helvetica Physica Acta 14,* 51–80 (1941), made use of a compensating scalar field; F. Bopp, "Eine lineare Theorie des Elektrons," *Annalen der Physik 38,* 345–384 (1940), and A. Landé, "Finite self-energies in radiation theory," *Physical Review 60,* 121–127 (1941), introduced vector fields for this purpose.

[18] V. Ambarzumian and D. Iwanenko, "Zur Frage nach Vermeidung der unendlichen Selbstrückwirkung des Elektrons, *Zeitschrift für Physik 64,* 563–567 (1930).

[19] G. Watagin, "Bemerkungen über die Selbstenergie der Elektronen," *Zeitschrift für Physik 88,* 92–98 (1934).

a "*G*-factor." In 1938 Heisenberg[20] emphasized the advantage of introducing a universal elementary length instead of a universal elementary mass. Schild,[21] in 1948, constructed a model of a (discrete) space-time discontinuum which admits a surprisingly large number of Lorentz transformations and even shows a number of striking resemblances to some of the features of Dirac's theory of the electron. Recently, theories of elementary particles have been formulated without using space-time coordinates at all, treating space-time instead as a statistical concept just as temperature is treated in the kinetic theory of gases.[22]

All in all, these corrective procedures, important as they were for the development of modern field theory, were mainly mathematical artifices which contributed little to a more profound understanding of the nature of mass. The same may be said of the so-called regularization procedures in quantum-field theory which were introduced for the first time by Kramers[23] in order to remove the infinities due to the electron interaction with the zero-point fluctuations of the electromagnetic field. The interaction contribution to the mass of the particle is simply ignored as inseparable from the total observable mass, which, being the sum of the self-mass and the mass that the particle would possess if stripped of its interaction with the field, is a finite quantity. The procedure of adjusting the constants to allow for a finite sum for the experimentally observable mass is generally referred to as renormalization. These renormalization procedures, apart from their problematic aspects from the purely mathe-

[20] W. Heisenberg, "Über die in der Theorie der Elementarteilchen auftretende universelle Länge," *Annalen der Physik 32,* 20–33 (1938).

[21] A. Schild, "Discrete space-time and integral Lorentz-transformations," *Physical Review 73,* 414–415 (1948). See also *Canadian Journal of Mathematics 1,* 29–47 (1949).

[22] Takao Tati, "A theory of elementary particles," *Progress in Theoretical Physics 18,* 235–246 (1957). See also *Nuovo cimento 4,* 75–87 (1956).

[23] H. A. Kramers, "Non-relativistic quantum-electrodynamics and correspondence principle," *Rapports du 8e Conseil Solvay 1948* (Brussels, 1950), pp. 241–265. Reprinted in *H. A. Kramers' collected scientific papers* (North Holland, Amsterdam, 1956).

matical point of view,[24] seem to lead neither to a consistent interpretation of the nature of mass nor to an unambiguous prediction of the mass spectrum of the elementary particles. A satisfactory quantum-electrodynamical explanation of the nature of mass is thus still a task for the future.

Another field theory which we have not mentioned as yet — and a field theory *par excellence* — is the general theory of relativity. It is therefore appropriate to conclude our investigation of the field-theoretical concept of mass with a few remarks on the contributions of general relativity toward the clarification of our concept.

Historically, mass appeared within the context of general relativity for the first time in connection with the so-called "principle of equivalence" announced by Einstein in 1916.[25] The equality or proportionality between inertial and gravitational mass indicated the possibility of "transforming away" homogeneous gravitational fields. Through the amalgamation of the gravitational force field into the space-time structure, the equivalence or proportionality between inertial and gravitational mass, which in Newtonian physics was an empirical and purely accidental feature, now becomes explicable as a consequence of the principle of covariance. This conclusion, important as it was as a unification of two originally different concepts, from the methodological point of view was based on the traditional notion of mass, or, in other words, did not yet call for a fundamental revision of the concept of mass as such (either inertial or gravitational).

It should, however, be recalled that the proportionality between inertial and passive gravitational mass — which in Newtonian physics was a matter of pure contingency — becomes a

[24] See, for instance, D. I. Blochincev, *Uspekhi fizicheskikh nauk 61,* 137 (1957), and *Fortschritte der Physik 6,* 246–269 (1958).

[25] Albert Einstein, "Die Grundlagen der allgemeinen Relativitätstheorie," *Annalen der Physik 49,* 769–822 (1916), translated into English as "The foundations of the general theory of relativity," in *The principle of relativity* (Dover, New York, 1923), pp. 109–164.

constitutive principle for the general theory of relativity. Let us clarify this point in more detail.

According to the principle of equivalence the so-called "inertial forces" (as, for instance, the centrifugal force), which in Newtonian mechanics bear the character of being fictitious owing to an inappropriate choice of the reference system, are interpreted in general relativity as real forces originating in the distant masses of the universe. Indeed, the rejection of a difference in principle between inertial and gravitational forces is the very essence of the principle of equivalence (only so-called nonpermanent gravitational fields are, of course, referred to). The centrifugal force,

$$F_c = m_i R\omega^2,$$

for instance, acting on a body of inertial mass m_i at a distance R from the origin of a coordinate system rotating with angular velocity ω must also be expressible, like any other gravitational force, as the product of a passive gravitational mass and the (negative) gradient of a potential Φ:

$$F_c = -m_p \operatorname{grad}_R \Phi = -m_p \frac{\partial \Phi}{\partial R},$$

which is the appropriate formulation of Poisson's equation for the case under consideration.

As can be easily seen from the expression of the line element ds^2 for the rotating reference system, the (scalar) potential Φ is given by the formula

$$\Phi = -\tfrac{1}{2}R^2\omega^2,$$

and consequently the centrifugal force is determined by the expression

$$F_c = m_p R\omega^2,$$

which shows that the identity $m_i = m_p$ lies at the foundations of the general theory of relativity.

In contrast to the fundamental identity of inertial and passive gravitational mass, general relativity cannot derive the identity of active and passive gravitational mass — which, as we have seen on page 126 was part and parcel of the foundations of Newtonian physics — from the action-reaction principle. For general relativity the action-reaction principle, based on the concept of action at a distance (and simultaneity) is incompatible with its field-theoretic approach. As we shall see, however, later on, other considerations will lead to the conclusion that in general relativity also a universal proportionality or equality between the two kinds of gravitational masses does exist.

The general relativistic concept of mass was at first in a situation similar to that which prevailed in classical quantum mechanics prior to the advent of quantum field theory. To make this point clear, consider the well-known example of Schwarzschild's line element for the case of a centrally symmetric and static gravitational field.[26] In order to arrive at a rigorous solution of Einstein's field equations for the case under discussion, Schwarzschild showed that the general expression

$$ds^2 = g_{mn}\, dx^m\, dx^n$$

of the space-time metric can be written under the above-mentioned restrictions as

$$ds^2 = g_{11}\, dr^2 - r^2(d\theta^2 + \sin^2\theta\, d\varphi^2) + g_{44}c^2\, dt^2,$$

and that the field equations impose upon g_{44} the condition

$$\frac{d}{dr} \log\,[r(g_{44} - 1)] = 0,$$

so that

$$g_{44} = 1 - \frac{2k}{r},$$

where k is an as yet undefined constant. Since a linear approximation of the field equations shows that $\frac{1}{2}c^2(g_{44} - 1)$ plays the role of the Newtonian potential Φ, it follows that

[26] K. Schwarzschild, "Über das Gravitationsfeld eines Massenpunktes," *Berliner Berichte* (1916), pp. 189–196.

$$g_{44} = 1 + 2\Phi.$$

Finally, since in Newtonian dynamics for the case under consideration Poisson's equation yields

$$\Phi = -G\frac{m}{r},$$

Schwarzschild is led to the conclusion that

$$g_{44} = 1 - \frac{G}{c^2}\frac{2m}{r}.$$

In other words, the constant k is identified by analogy conclusions with the active gravitational mass m of the central body (the proportionality factor G/c^2 leads merely to a change in units). The rate at which the metric deviates from flatness is thus interpreted as mass.

Now it will be clear why the early development of general relativity, as far as the concept of mass is concerned, shows a similarity to the early stages of quantum mechanics. In either theory the concept of mass has been introduced by analogy, and in either theory it had to be introduced in order to provide the otherwise abstract theory with points of contact with empirical data and facts. But in both theories, at that stage of their development, the notion of mass was an illegitimate element, foreign to their conceptual texture. The reason for this illegitimacy in general relativity is, of course, of quite a different character from that outlined above with regard to quantum mechanics. In general relativity mass or its equivalent energy is in general not a component of a tensor, in contrast to density, which is the T_{44} component of the momentum-energy tensor. Since the tensor T_{mn} in general relativity *qua* field theory determines the behavior of physical processes or events, it would be logical to define mass as the integral of T_{44} over three-dimensional space. This integral, however, is a tensor component only in a space of zero curvature. Thus the classical definition of mass as volume times density (Newton's Definition 1), which conceptually harmonizes with the field-theoretical point of view, becomes unacceptable and a more general concept of mass has to be adopted.

The problem of how to define the mass (or energy) of a dynamical system in general relativity in an unambiguous, covariant, and physically meaningful way was given much attention by Einstein,[27] Nordström,[28] Klein,[29] Weyl,[30] and others. In search for a solution it seemed natural to generalize the procedure adopted for the special theory of relativity. It will be recalled that in the special theory the momentum-energy four-vector P (page 164), whose time-component is $-E/c$, satisfies the relation

$$P^2 - E^2/c^2 = -m_0^2c^2, \tag{16}$$

where P^2 is the sum of the squares of the spatial components of P and m_0 is the inertial proper mass of the particle or system under discussion. Now, Eq. (16) itself may be regarded as a definition of the mass, provided the terms in the left-hand member of the equation can be determined independently.

The question thus arose whether in general relativity a momentum-energy vector for a given dynamical system exists. The problem was to some extent solved by Einstein and Klein.[31] In the special theory, as we know, the conservation laws of momentum and energy are expressible by the Lorentz invariant differential equation

$$\operatorname{div} T = \frac{\partial T_{ik}}{\partial x^k} = 0, \tag{17}$$

where T_{ik} is the total momentum-energy tensor of the system.

[27] A. Einstein, "Der Energiesatz in der allgemeinen Relativitätstheorie," *Sitzungsberichte der Preussischen Akademie der Wissenschaften* (1918), part 1, pp. 448–459.

[28] G. Nordström. "On the mass of a material system according to the theory of Einstein," *Proceedings of the Koninklijke Akademie van Wettenschappen te Amsterdam 20,* No. 7 (1917).

[29] F. Klein, "Über die Integralform der Erhaltungssätze und die Theorie der räumlich geschlossenen Welt," *Göttinger Nachrichten* (1918), pp. 394–423.

[30] H. Weyl, *Space-time-matter* (Methuen, London, 1922), pp. 268–273.

[31] References 27 and 29.

The corresponding formulation for general relativity would require that the covariant divergence of the momentum-energy tensor vanish. The vanishing of a covariant divergence of a tensor of the second rank — in contrast to that of a vector, such as the current–charge-density vector in general relativistic electrodynamics[32] however, does not entail the vanishing of the ordinary divergence, as required for conservation. Nevertheless, Einstein showed that the laws of conservation could be written in the form

$$\frac{\partial \mathfrak{T}_i^k}{\partial x^k} = 0$$

with $\mathfrak{T}_i^{\ k} = (-g)^{1/2}(T_i^{\ k} + t_i^{\ k})$, where the $(-g)^{1/2} t_i^{\ k}$, which Einstein called "the components of energy of the gravitational field," are built up of the g^{mn} and their first derivations.

By the usual application of the four-dimensional Gauss theorem it can be shown that the quantities

$$P_i' = \frac{1}{c} \int \mathfrak{T}_i^4 \, dx^1 \, dx^2 \, dx^3 \quad (i = 1, 2, 3, 4)$$

are constant in time. In addition, Klein demonstrated that the P_i' behave like vectors under linear transformations. Since the P_i' reduce in space of zero curvature to the P_i of special relativity, it was natural to define the mass of the system, in analogy to Eq. (16), by the equation

[32] Let it be recalled that for a contravariant vector φ^i

$$\text{div } \varphi^i = \varphi^i_{\ ,i} = \frac{1}{(-g)^{1/2}} \frac{\partial}{\partial x^i} ((-g)^{1/2} \varphi^i) = 0$$

implies, of course,

$$\frac{\partial}{\partial x^i} [(-g)^{1/2} \varphi^i] = 0,$$

whereas for a tensor of the second rank the equation

$$\text{div } T^{ik} = T^{ik}_{\ \ ;k} = \frac{\partial T^{ik}}{\partial x^k} + \Gamma^i_{kr} T^{rk} + \Gamma^k_{kr} T^{ir} = 0$$

does not imply

$$\frac{\partial T^{ik}}{\partial x^k} = 0.$$

$$m_0 = \frac{1}{c}\,(P_4'^2 - P_1'^2 - P_2'^2 - P_3'^2)^{1/2}. \qquad (18)$$

Such a definition of mass would make sense in the general theory of relativity only if the P_i' were independent of the choice of the coordinate system. Unfortunately, as long as Einstein's "components of energy of the gravitational field" are employed, they are not. Einstein could merely show that the P_i' are independent of the choice of quasi-Galilean coordinate systems, that is, coordinates which at sufficiently large spatial distances from the system and from the four-space region traversed by it (its so-called "canal" or "tube") assume the metric of Minkowski space.

In view of these difficulties, Eddington and Clark,[33] in 1938, suggested that the mass of a dynamical system be defined as the "mass M of an equivalent particle which gives the same line-element at great distances." More precisely, it is the mass of the system at the instant considered while the distances have to be not so large as to make the time lag of the potentials physically significant. Furthermore, the velocity and the acceleration of the comparison particle have to be equal to those of the center of mass[34] of the system.

It is clear that the present concept of mass of a gravitating mechanical system takes account of the fact that in accordance with the equivalence of mass and energy any potential or kinetic energy residing in the system contributes to its "mass." In fact, Gilloch and McCrea,[35] who employed this definition in their calculation of the mass of a rotating cylinder, showed that this "mass" is equal to the proper mass of the cylinder plus its kinetic energy divided by c^2. In general, however, the mass of a system,

[33] Sir Arthur Eddington and G. L. Clark, "The problem of n bodies in general relativity theory," *Proceedings of the Royal Society 166*, 465–475 (1938).

[34] The concept "center of masses" is geometrically definable; see reference 33, p. 468.

[35] Josephine M. Gilloch and W. H. McCrea, "The relativistic mass of a rotating cylinder," *Proceedings of the Cambridge Philosophical Society 47*, 190–195 (1951).

as Eddington and Clark themselves pointed out,[36] is equal to the sum of the proper masses and the energies of the bodies comprising the system only if the moment of inertia C (about the center of mass) is unaccelerated. If m_i are the proper masses of the system, and K and V the kinetic and potential energies, respectively, the general formula for the "mass" of the system is

$$M = \sum_i m_i + \frac{1}{c^2}(K + V) + \frac{1}{2}\frac{d^2C}{dt^2}. \qquad (19)$$

The classical principle of the additivity of mass is, of course, no longer valid. The prerelativistic possibility of tagging separate constituents of a dynamical system with individual mass values seems no longer justified. This is, of course, the price field theory had to pay for the emancipation of the concept of energy, as explained in Chapter 13. In fact, using sociological terminology, one is tempted to say that the "emancipation of energy" led to the "collectivism of mass."

The Eddington-Clark definition of mass, as Eddington himself pointed out only two years after the publication, is not yet satisfactory from the general relativistic point of view. For it can be shown that the quantity defined is the space integral of $T_{44} + t_{44}$, where t_{44} is an expression for the potential energy. The terms t_{mn}, however, the so-called energy-momentum pseudotensor, are algebraic functions of the gravitational-field intensities (the first derivatives of the g_{mn}) and consequently, in the case of nonlinear coordinate transformations, are not tensors.

Clark, in an article "On the gravitational mass of a system of particles,"[37] reexamined certain suggestions advanced in 1935 by Whittaker[38] and Ruse,[39] according to whom the notion of mass in general relativity can be defined by means of Gauss's theorem in its four-dimensional formulation. In analogy to electrostatics,

[36] Reference 33, p. 469.

[37] *Proceedings of the Royal Society of Edinburgh 62*, 412–423 (1949).

[38] E. T. Whittaker, "On Gauss' theorem and the concept of mass in general relativity," *Proceedings of the Royal Society 149*, 384–395 (1935).

[39] H. S. Ruse, "Gauss' theorem in a general space-time," *Proceedings of the Edinburgh Mathematical Society 4*, 144–158 (1935).

where the charge of a system is determined by the total flux of the electrostatic field vector through a closed surface surrounding the system, the total flux of the relativistic gravitational force through a simple closed surface is proportional to the total gravitating mass enclosed, the factor of proportionality being -4π. The relativistic treatment is, however, complicated by the fact that the gravitational force, as measured by any observer, is a function of the observer's position as well as of his velocity and acceleration. Whittaker succeeded in generalizing Gauss's theorem for the case under discussion for a static gravitational field, and Ruse, in his turn, generalized Whittaker's formula for a general metric of the space-time $ds^2 = g_{mn}\, dx^m\, dx^n$. The role of the charge in Gauss's theorem in electrostatics is taken over by $T_{44} - \frac{1}{2}T$ in Whittaker's computation, and by $\lambda^i\lambda^k T_{ik} - \frac{1}{2}T$ in Ruse's, where T_{ik} is the energy tensor, $T = g^{mn}T_{mn}$, and λ^i are the directions (unit vectors) of the world-lines of the fundamental observers. The Newtonian concept of mass is thus consistently replaced by that of the energy tensor. Since the latter does not necessarily vanish in the absence of matter in the classical sense of the word, but is different from zero in the presence of fields, such as the electrostatic field, gravitating mass has naturally to be ascribed also to any form of energy contained in a field of forces.

From a rigorous and consistent field-theoretical point of view, the representation of matter and energy by means of the tensor T_{mn} must be considered only a provisional device which eventually has to be replaced by purely field-theoretic methods. Einstein himself very much disliked "the illegal marriage between the artificial energy-momentum tensor of matter and the curvature tensor." [40] The Whittaker-Ruse-Clark approach to the concept of mass thus seemed not yet to be the final answer to our problem. But before we discuss the concept of mass from the viewpoint of

[40] See L. Infeld, "On equations of motion in general relativity theory," *Helvetica Physica Acta, Supplementum IV, Jubilee of Relativity Theory, Proceedings* (Birkhäuser Verlag, Basel, 1956), p. 207. See also A. Einstein, *The meaning of relativity* (Princeton University Press, Princeton, ed. 4, 1953), pp. 81, 106, 165.

those theories that regard matter as singularities of the field, let us prove the proportionality of active gravitational and inertial mass on the basis of the energy-momentum tensor.

Einstein's original formulation of the energy or mass of a closed system, as we saw on page 209, was based on the use of quasi-Galilean coordinates and — what was an even more serious defect — did not make the notion of energy distribution in space physically meaningful.[41]

In an important article "On the localization of the energy of a physical system in the general theory of relativity,"[42] Møller recently showed that if Einstein's total canonical energy-momentum pseudotensor $\mathfrak{T}_i^k$, which can be expressed also[43] as the divergence of

$$h_i^{kj} = \frac{g_{in}}{2\kappa(-g)^{1/2}} \frac{\partial}{\partial x^m} [(-g)(g^{kn}g^{jm} - g^{jn}g^{km})] \qquad (20)$$

(κ is the Einstein constant of the field equations, related to the constant of gravitation G by the equation $\kappa = 8\pi G/c^4$), is replaced by a new momentum-energy pseudotensor $h_i^{kj} + \psi_i^{kj}$, where ψ_i^{kj} is defined by

$$\psi_i^{kj} = h_i^{kj} - \delta_i^k h_r^{rj} + \delta_i^j h_r^{rk} \qquad (21)$$

with a vanishing divergence, then both above-mentioned deficiencies can be remedied. The Einstein-Klein approach to the general relativistic definition of inertial mass for a closed dynamical system by Eq. (18), where m_0 was the inertial mass m_i, thus regained considerable interest.

An important conclusion concerning the relation between active gravitational mass and inertial mass in general relativity can be drawn from Eq. (21), or from Eq. (20), if for the sake of simplicity we return to the use of quasi-Galilean coordinates.

In the case of a static system, sufficiently removed from any

[41] The integrand of the space integral for the total energy is not invariant even for purely spatial transformations; see p. 210.

[42] *Annals of Physics 4,* 347–371 (1958).

[43] See, for instance, reference 6, p. 317.

other matter in the universe, the metric at sufficiently large spatial distances is of the Schwarzschild type:[44]

$$ds^2 = -\frac{1}{1 - 2k/r}\, dr^2 - r^2(\sin^2\theta\, d\varphi^2 + d\theta^2) + \left(1 - \frac{2k}{r}\right) c^2\, dt^2. \quad (22)$$

In isotropic coordinates this line element, for large r, can be written

$$ds^2 = -\left(1 + \frac{2k}{r}\right)(dx^2 + dy^2 + dz^2) + \left(1 - \frac{2k}{r}\right) c^2\, dt^2, \quad (23)$$

so that

$$g_{11} = g_{22} = g_{33} = -\left(1 + \frac{2k}{r}\right), \quad g_{44} = 1 - \frac{2k}{r},$$

and all the other g_{ik} vanish. Calculating $(-g)^{1/2}$ to the first order in r^{-1}, we obtain

$$(-g)^{1/2} = 1 + \frac{2k}{r}, \quad g^{11} = g^{22} = g^{33} = -\left(1 - \frac{2k}{r}\right), \quad g^{44} = 1 + \frac{2k}{r},$$

and all the other g^{ik} vanish.

For a static system the spatial components of P_i' vanish and according to Eq. (18)

$$E/c^2 = m_i \quad (24)$$

because $-cP_4' = E$. But E, as the time component of P_i', can also be computed in terms of the pseudotensor h_i^{kj} in accordance with the general formula

$$P_i' = \frac{1}{c}\int \mathfrak{T}_i^4\, dx^1\, dx^2\, dx^3 = \frac{1}{c}\int h_{i,j}^{4j}\, dx^1\, dx^2\, dx^3.$$

For $i = 4$ we obtain

$$E = -\int \frac{\partial}{\partial x^j}(h_4^{4j})\, dx^1\, dx^2\, dx^3,$$

or, applying Gauss's theorem,

$$E = -\int h_4^{4j}\gamma^{-1/2}\, d\sigma_j, \quad (25)$$

[44] *Ibid.*, p. 307.

where γ is the determinant of the spatial metric tensor $\gamma_{ik} = g_{ik}$ $(i, k = 1, 2, 3)$, since all g_{14} for a static system vanish $(i = 4)$; $d\sigma_j$ is the pseudovector normal to the surface element $d\sigma$ of the sphere S over which the integration in Eq. (25) has to be performed.

Substituting the values of $(-g)^{1/2}$, g_{ik}, g^{ik} in Eq. (20) we obtain for $j = 1, 2, 3$

$$h_4^{4j} = \frac{g_{44}}{2\kappa(-g)^{1/2}} \frac{\partial}{\partial x^j} [(-g)(g^{44} g^{jj})]$$

$$= -\frac{2k}{\kappa r^3} x^j.$$

Consequently,

$$E = \int \frac{2k}{\kappa r^3} x^j \gamma^{-1/2} \, d\sigma_j,$$

and, as the radius of the sphere increases to infinity,

$$E = \lim_{r \to \infty} \left[\frac{2k}{\kappa r^2} 4\pi r^2 \right] = \frac{8\pi k}{\kappa}. \tag{26}$$

Now it will be recalled that the active gravitational mass m_a was defined (page 206) by means of the Schwarzschild constant k according to the relation $k = (G/c^2) m_a$. Substituting this value for k in Eq. (26) and taking account of $\kappa = 8\pi G/c^4$, we finally get

$$E/c^2 = m_a. \tag{27}$$

Comparing this result with Eq. (24) we see that in virtue of the above mentioned analogy procedure the active gravitational mass of a body or dynamical system is equal to its inertial mass also in the general theory of relativity.

On page 204 we showed that the identity of inertial mass and passive gravitational mass is a direct consequence of the principle of equivalence. In view of our present result we thus come to the important conclusion that general relativity, unlike classical mechanics, considers the identity of *all three* kinds of masses a necessary — and not only accidental — constituent in its logical structure.

In general relativity a particle can also be regarded as a singularity in the field g_{mn}. Confining our discussion to empty space, we note that, in view of the fact that the field equations $G_{mn} = 0$ have to be satisfied everywhere outside the singularities, these latter, that is, the four-dimensional world-lines of the particles, cannot be arbitrarily specified. The mathematical restrictions imposed by the field equations upon the singularity curves in four-dimensional space are an expression of the fact that the laws of motion in general relativity are not superadded conditions that have to be complied with — as in Newtonian physics — but are rather direct consequences of the field equations themselves.[45] Since the notion of mass is first and foremost that of a motion-determining factor, and since the field is the primary and ultimate datum, the only logically as well as methodologically satisfactory way of introducing the concept of mass is via the derivation of the laws of motion from the field equations. The problem of determining the laws of motion from the field equations alone was solved in 1938 by Einstein, Infeld, and Hoffmann and a simplified solution of the problem was published two years later by Einstein and Infeld.[46] Finally, a logically still simpler but technically more tedious solution, accessible to approximations of an arbitrarily high order, was given by Einstein and Infeld in 1949.[47] It is probably one of the most profound and far-reaching treatments of this subject. Let us see how it introduces the concept of mass into the conceptual structure of its exposition.

The gravitational potentials g_{mn} are represented as $g_{mn} =$

[45] A. Einstein and J. Grommer, "Relativitätstheorie und Bewegungsgesetz," *Sitzungsberichte der Preussischen Akademie der Wissenschaften* (1927), pp. 2–13, 235–245.

[46] A. Einstein, L. Infeld and B. Hoffmann, "Gravitational equations and the problem of motion," *Annals of Mathematics 39*, 65–100 (1938); A. Einstein and L. Infeld, "Gravitational equations and the problem of motion II," *ibid. 41*, 455–464 (1940).

[47] A. Einstein and L. Infeld, "On the motion of particles in general relativity theory," *Canadian Journal of Mathematics 1*, 209–241 (1949). This is only a summary of the results. The manuscript containing the calculations in full has been deposited at the Institute for Advanced Study, Princeton, New Jersey. In our résumé we use Latin indices for summation over 1 to 4.

$\eta_{mn} + h_{mn}$, where η_{mn} are the Galilean values and h_{mn} the (not necessarily small) deviations of space-time from flat space. For convenience certain linear combinations of the h_{mn} are defined as $\gamma_{mn} = h_{mn} - \frac{1}{2}\eta_{mn}\eta^{pq}h_{pq}$. The first (nonzero) term in the expansion of γ_{44} is essentially a term denoted by $\underset{2}{\gamma_{44}}$. The first-order approximation of the field equations then shows that $\underset{2}{\gamma_{44,ss}} = 0$. Now, it can be shown that this last equation is analogous to Poisson's equation and its solution can be written as $\underset{2}{\gamma_{44}} = 2\varphi$ with $\varphi = \sum_{s=1}^{N} \left(- 2\underset{2}{\overset{s}{m}}\overset{s}{\psi} \right)$, where the summation extends over the N singularities present in the field; $\overset{s}{\psi}$ is the reciprocal of the spatial distances from the sth singularity and the $\underset{2}{\overset{s}{m}}$ are time-independent positive constants as yet undetermined.

In order to identify the physical significance of these constants, Einstein and Infeld calculate the field (associated with the first "particle") when all the other singularities are removed from the first "particle" characterized by the constant $\underset{2}{\overset{1}{m}}$. By comparing the result with the gravitational field produced by an isolated body of gravitational mass M as computed by "conventional methods in general relativity," such as the Schwarzschild line element, the authors find that for large r this latter field proves to be identical with the first-order-approximation field, provided $M = \underset{2}{\overset{1}{m}}$. Consequently, "$\underset{2}{\overset{1}{m}}$ is the gravitational mass, since for large r the field is that of a particle with gravitational mass $\underset{2}{\overset{1}{m}}$."

Thus, a recourse to analogy (in the present case even by analogy to an analogy) to classical dynamics renders those constants of integration physically meaningful and designates them as "masses." But, strictly speaking, is it really necessary, within the context of a rigorous field theory, to search for a dynamical interpretation of these constants of integration? Does not their purely mathematical aspect as constants in the space-time functions of the four-dimensional curves constitute their full meaning? The answer would be in the affirmative were their intrinsic numerical values irrelevant. These values are important, however, for the actual determination of the four-dimensional trajectories of the singularities. These values can be inferred only through the configuration pattern of at least two such singularities, or, in other words, by the study of their interaction. But this means that a dynamical interpretation, in the traditional sense, is a requirement for the description of their motion. In short, a concession to traditional concepts has to be made even in such an advanced treatment in modern field theory.

Defining mass as a singularity in the metric, however, raises additional problems. First of all, there is the question of principle as to what kind of singularities are admissible. As Wheeler recently stated,[48] "there does not exist today even the beginnings of a comprehensive analysis of the kinds of singularities which may arise in solutions of Einstein's field equations." Secondly, "mass" ceases to be a well-defined quantity in the case of several singularities as soon as the constituent parts of a system are separated from each other by distances of the order of their own gravitational radii. In fact, the very concept of "number of singularities" seems to defy definability. It will therefore be understood that even the Einstein-Infeld analogy approach cannot be considered as leading to an unambiguous definition of mass.

In conclusion of this chapter on the concept of mass in field theory, the so-called space-theories of matter have to be considered. In their attempt to reduce physics to the geometry of space, the problem of mass is, of course, of paramount importance for these theories.

If we ignore prescientific space theories of matter, such as

[48] John A. Wheeler, "Geometrodynamics and the problem of motion," *Reviews of Modern Physics 33,* 64 (1961).

certain ancient Vedic concepts[49] based on a belief in the insubstantiality of the phenomenal world, or certain Pythagorean and Platonic teachings[50] and similar speculations, one of the more important attempts in modern times was the theory of William Kingdon Clifford, the English translator of Riemann's mathematical writings on the structure of space. Clifford, it will be recalled, conceived of matter and its motion as a manifestation of the varying curvature of space. In 1876 Clifford published an essay "On the space theory of matter" (an enlarged version of a paper presented to the Cambridge Philosophical Society) in which he asserts the ultimate identity of space and matter. Space, in his view, is not merely the arena for the occurrence of physical events; rather it is the ultimate and exclusive building material of physical reality. "In the physical world nothing else takes place but this variation [of the curvature of space]." [51] Clifford, however, was not in a position to carry out his ambitious program and, in particular, could not interpret the notion of mass in terms of purely spatial or geometric considerations.

Meanwhile the problem of an intrinsic relation between the structure of space and the laws of dynamics and electrodynamics engaged the attention of philosophers and physicists alike. In fact, Kant, in his precritical *Thoughts on the true estimation of living forces*[52] believed in such a connection and tried to derive the three-dimensionality of space from Newtonian dynamics: "It is probable that the three-fold dimensionality of space is a consequence of the law according to which the forces of sub-

[49] See the theory of ākāśa in the *Brihadāranyaka Upanishad,* part 2, chap. 3, secs. 2 and 3; in the *Taittiriya Upanishad,* part 2, chap. 1, sec. 3; in the *Chhāndogya Upanishad,* part 1, chap. 9, sec. 1, part 3, chap. 18, sec. 6.

[50] See Max Jammer, *Concepts of space* (Harvard University Press, Cambridge, 1952), p. 12.

[51] W. K. Clifford, *The common sense of the exact sciences* (ed. J. R. Newman; Knopf, New York, 1946), p. 202.

[52] *Gedanken von der wahren Schätzung der lebendigen Kräfte,* secs. 10–11. English translation in John Handyside, *Kant's inaugural dissertation and early writings on space* (Chicago, 1929).

stances act on each other."[53] In 1808 Laplace tried to show that the postulation of scale invariance of linear extensions in the physical universe implies the specific formulation of Newton's inverse-square law, any exponent other than 2 being incompatible with the assumption. Other investigations of this kind on the relation between metric and dynamics were carried out by Delboeuf,[54] Bertrand,[55] and Zenneck.[56] Einstein's field equations of general relativity, according to which the fundamental metric tensor g_{mn} depends upon the mass-energy tensor T_{mn}, seemed to give a clear-cut answer to the problem under discussion as far as (mechanical) dynamics is concerned: geometry became a branch of physics and space a physical object.

With respect to electrodynamics, however, a different development took place. Vito Volterra,[57] as early as 1889, realized that Maxwell's equations constitute merely a special case of a general theory of conjugate functions. Friedrich Kottler, in 1922, published two interesting papers[58] in which he tried to demonstrate that, just as Newton's laws and the geometry of space have no necessary intrinsic connection,[59] so also the field physics of Maxwell's theory is independent of any metric. Kottler thus initiated a movement that aimed at a complete elimination of any metrical relations, either Riemannian or conformal, from the fundamental

[53] "Es ist wahrscheinlich, dass die dreifache Abmessung des Raumes von dem Gesetze herrühre, nach welchem die Kräfte der Substanzen in einander wirken . . . Die dreifache Abmessung scheint daher zu rühren, weil die Substanzen in der existierenden Welt so ineinander wirken, dass die Stärke der Wirkung sich wie das Quadrat der Weiten umgekehrt verhält."

[54] See L. Couturat, "Note sur la géométrie non-euclidienne et la relativité de l'espace," *Revue de métaphysique et de morale 1,* 302 (1893).

[55] J. Bertrand, *Comptes rendus 77,* 846 (1873).

[56] J. Zenneck, "Gravitation," in *Enzyklopädie der mathematischen Wissenschaften* (Teubner, Leipzig, 1903–1921), vol. 5, part 2, p. 42.

[57] Vito Volterra, "Sulle funzioni coniugate," *Rend. Accad. dei Lincei 5,* 599–611 (1889); reprinted in V. Volterra, *Opere matematiche* (Accademia Nazionale dei Lincei, Rome, 1954), vol. 1 (1881–1892), pp. 420–432.

[58] F. Kottler, "Newton's Gesetz und Metrik," *Wiener Sitzungsberichte 131,* 1–14 (1922); and "Maxwell'sche Gleichungen und Metrik," *ibid.,* 119–146.

[59] "Das Newtonsche Gesetz und die Geometrie unseres Raumes stehen in keinem notwendigen Zusammenhang."

laws of physics. More recently D. van Dantzig became one of the most eloquent proponents of this movement. It is well known that Maxwell's equations are invariant under orthogonal transformations in three-dimensional space, a fact warranted by their usual vector representation itself. Maxwell's equations are invariant also with respect to the more general group of affine transformations, as can easily be shown by Klein's principle of adjunction. Their invariance with respect to Lorentz transformations is one of the fundamental issues of Einstein's theory of relativity. That they are invariant under a much wider group of transformations, the so-called conformal transformations, has been shown by Cunningham[60] and by Bateman.[61] Finally, that Maxwell's equations are totally independent of any metric whatever has been shown by van Dantzig.[62]

Simultaneously, however, the diametrically opposite movement toward a unification of gravitation and electromagnetism on the basis of a suitable metric achieved important results.[63]

It was within this context that G. Y. Rainich, in 1925, published an interesting study on the relation between the Riemann curvature tensor and the electromagnetic field tensor. In his paper "Electrodynamics in the general relativity theory,"[64] the importance of which was not understood — and could not be understood — at the time, Rainich, in contrast to Kottler's program, showed "that under certain assumptions the electromagnetic field is entirely determined by the curvature of space-time, so that there is no need of further generalizing the general rela-

[60] "The principle of relativity in electrodynamics and an extension thereof," *Proceedings of the London Mathematical Society 8,* 77–98 (1910).

[61] "The transformation of the electrodynamical equations," *ibid.,* 223–264.

[62] "The fundamental equations of electromagnetism, independent of metrical geometry," *Proceedings of the Cambridge Philosophical Society 30,* 421–427 (1934). See also D. van Dantzig, "Electromagnetism, independent of metrical geometry," *Proceedings of the Koninklijke Akademie van Wetenschappen te Amsterdam 37,* 521–525 (1. The foundations), 526–531, 643–652 (1934); *39,* 126–131 (1936).

[63] A. Einstein, W. Mayer, T. Kaluza, O. Klein, H. Weyl, O. Veblen, B. Hoffmann, to mention only a few of its proponents.

[64] *Transactions of the American Mathematical Society 27,* 106–136 (1925).

tivity theory."[65] In fact, Rainich showed that a Riemannian space with a non-null differentiable Ricci curvature tensor R_{mn} of zero trace $R_n{}^n = 0$, whose square is a certain multiple of the unit matrix

$$R^m_a R^b_m = \delta^b_a(\tfrac{1}{4}R_{st}R^{st})$$

and in which the vector

$$A_b = (-g)^{1/2}\, \epsilon_{bkmn} R^{kp;m} R^n_p / R_{st}R^{st}$$

(ϵ_{bkmn} is the completely antisymmetric pseudotensor of rank 4) satisfies the condition

$$A_{b;p} - A_{p;b} = 0,$$

describes without further assumptions Maxwell's source-free electrodynamics. Thus, under certain conditions (Rainich's field conditions), the geometry of space alone (the contracted curvature tensor) determines the local values of the electromagnetic field tensor and Maxwell's equations are merely geometric statements connecting the Ricci curvature and its rate of change. The importance of Rainich's results for a deductive construction of a space theory of matter was not realized until Charles W. Misner[66] arrived independently at the same conclusions. The possibility of expressing the relativistic formulas of Maxwell's equations in a purely geometric form opened up a new approach to a consistent space theory of matter. Abandoning the tacit assumption made hitherto that space is simply connected, Wheeler and Misner demonstrated the compatibility of Riemannian geometry with a large range of multiply connected topologies and showed that certain appropriately chosen topologic linkages simulate electric charges in the sense that they are externally indistinguishable from ordinary electric charges subject to mutual attractive or repulsive forces, Gauss's theorem, and the law of conservation of charge.

[65] *Ibid.*, 107.

[66] Charles W. Misner and John A. Wheeler, "Classical physics as geometry," *Annals of Physics 2*, 525–603 (1957).

Having established that Maxwell's electrodynamics is but a manifestation of geometric (topologic) properties and that charges are expressible in terms of source-free electromagnetic fields, Wheeler and Misner attempted to derive the concept of mass also in terms of geometric characteristics. Strictly speaking, part of the solution was already at hand. For an electromagnetic field carries energy density. But in order to deduce mass as that of a physical body (an object possessing mass and position coordinates) the energy has to be localized and the electromagnetic field itself has to form a relatively stable and concentrated entity.

That this possibility can actually occur became clear when Wheeler, in 1955, demonstrated the existence of certain nonsingular solutions of the coupled equations of relativity and electromagnetism. He showed[67] that Einstein's field equations

$$R_{mn} - \frac{1}{2} g_{mn} R = \frac{8\pi G}{c^4} T_{mn},$$

coupled with Maxwell's equations

$$(-g)^{-1/2} \frac{\partial}{\partial x^k} [(-g)^{1/2} F^{ik}] = 0,$$

where $F_{ik} = \partial A_k / \partial x^i - \partial A_i / \partial x^k$ (A_k are the electromagnetic potentials) and where $T_i^k = (1/4\pi)(F_{ir}F^{kr} - \frac{1}{4} F_{rm}F^{rm}\delta_i^k)$ is the expression of the electromagnetic energy-momentum tensor, admit of a completely nonsingular solution for the A_k and g_{mn}. As the equations indicate, the gravitational mass originates solely from the energy stored in the electromagnetic field. Thus, the gravitational attraction exerted by the energy of the electromagnetic disturbance is capable of concentrating the disturbance and holding it together for a time long in comparison with the characteristic periods of the system. A simple variety of such conglomerations of electromagnetic energy, or "geons" (gravitational-

[67] John Archibald Wheeler, "Geons," *Physical Review 97*, 511–536 (1955); see also Edwin A. Power and John A. Wheeler, "Thermal geons," *Reviews of Modern Physics 29*, 480–495 (1957).

electromagnetic entities), has a circular toroidal shape. The existence of solutions of the coupled equations corresponding to other configurations of energy concentration is a subject of contemporary research.[68] The geon concept thus provides a field-theoretic representation of what classical physics regarded as a physical body, possessed of mass (inertia) and localization in space (position coordinates).

Combining Rainich's space theory of source-free electrodynamics with the concept of geons, we come to the conclusion that "in geometrodynamics mass and charge . . . are aspects of the geometrical structure of space."[69] Geometrodynamics thus seems to have found a memorable roundabout way of basing the concept of mass solely on the geometric notion of curved space. For an assessment of this theory, however, the difficulties and deficiencies should not be ignored. Only masses of the order of 10^{39} gm to 10^{57} gm can be accounted for, masses not found in nature. Any attempt to modify the theory into a "quantum geometrodynamics"[70] for microphysics and to interpret the masses of elementary particles as corresponding to characteristic states of excitation of collective disturbances in the metric still faces serious difficulties. In short, although a new and interesting vista has been opened up toward a geometric interpretation of the concept of mass, for the time being "real masses," as considered in physics, cannot yet be accounted for.

This, then, is the story of the basic concept of mass.

Although of primary importance in all branches of physics and an indispensable conceptual tool of scientific thought, the notion

[68] See, for instance, F. J. Ernst, Jr., "Linear and toroidal geons," *Physical Review 105,* 1665–1670 (1957); D. R. Brill and J. A. Wheeler, "Interaction of neutrinos and gravitational fields," *Reviews of Modern Physics 29,* 465–479 (1957), L. G. Chambers, "The Hund gravitational equations and geons," *Canadian Journal of Physics 37,* 1008–1016 (1959).

[69] Reference 66, p. 595.

[70] John A. Wheeler, "On the nature of quantum geometrodynamics," *Annals of Physics 2,* 604–614 (1957).

of mass seems to elude all attempts at a fully comprehensive elucidation and a logically as well as scientifically unobjectionable definition.

Throughout its long history in human thought, from its early adumbrations in Neo-Platonic philosophy, its mystic and still inarticulate presentation in theology, to its scientific manifestation in the physics of Kepler and Newton, to its carefully thought-out redefinitions in positivistic and axiomatic formulations, up to its far-reaching modifications in modern theories of physics — nowhere does science seem to get full command and control over all the conceptual intricacies involved. One has to admit that in spite of the concerted effort of physicists and philosophers, mathematicians and logicians, no final clarification of the concept of mass has been reached.

The modern physicist may rightfully be proud of his spectacular achievements in science and technology. However, he should always be aware that the foundations of his imposing edifice, the basic notions of his discipline, such as the concept of mass, are entangled with serious uncertainties and perplexing difficulties that have as yet not been resolved.

INDEX